Manufactured Genetic Origins: The Fake Eurasian Back Migration

Dr. CLYDE WINTERS

DEDICATION

This book is dedicated to Everyone who seeks Truth.

Table of Contents

INTRODUCTION

Geneticists claim that the ancestors of modern Europeans migrated from the Middle East to Europe. Although archaeologists have accurately documented the spread of Bell Beaker and Yamnaya cultures from the Levant and Anatolia, into Europe they have failed to accurately designate the ethnicity of the people who carried these cultures to Europe.

These geneticist maintain that the first Neolithic migrants to Europe were Indo-European speakers and therefore, the ancestors of contemporary Europeans .This is a manufactured origin for Europeans because, it is not founded on historical and archaeological evidence. Although, this is their view , the archaeological evidence makes it clear that the founders of civilization in the Middle East came directly from Africa. In the ancient literature these people were called Kushites not Indo-Europeans (Winters, 2018). This population based on the skeletal remains , were negro or African people-not Caucasian.

Given the fact that the carriers of Bell Beaker and Yamnaya

cultures were Africans, instead of contemporary Europeans begs the question is the proposed genetic origin of contemporary Europeans in the Middle East, as manufactured origin for Europeans based on racism, White Supremacy and a non-existent back migration of Eurasians into Africa.

In this book I will answer the Question: "Is the alleged beginning of contemporary Europeans in the Middle East manufactured to give Europeans greater self-esteem?" Secondly, the Eurasian back migration to Africa, is a fake historical event. A fake historical event to white out Black and African people out of ancient World History.

Many archaeologists and genetics believe that due to the success of using DNA evidence in the courtroom, DNA is ultimate proof that identifies this or that population that represented that was associated with a particular culture or civilization based solely on genome, recovered from ancient bones; and contemporary populations analyzed by Bayesian statistics.

There are two major problems associated with population genetics. First, genetics appears and " to be a pseudoscience because, it does not use the scientific method to accumulate knowledge It uses expost facto or descriptive research methods, which can only describe the presence of a genome. It cannot really specify the origin of the genome and "population" that carried the genome.

A Descriptive study collects data to test a hypothesis or questions related to the current status of the subject of the study.

The second problem with population genetics is the foundation of population genetics theory, that the varied human populations or races were separated until the Atlantic Slave Trade, sent millions of Africans into bondage in the Americas and Eurasia. David Reich (2018) wrote that "the ancestors of East Asians, Europeans, West Africans, and Australians until recently were, almost completely isolated from one another for 40,000 years or longer…"

This theory is groundless Africans and Eurasians had been mixing in the Americas and Eurasia for thousands of years prior to 1492. Because the admixture of Africans and Eurasians is thousands of years earlier than 1492, makes the baseline used to determine population migration lacks any validity and reliability

Research Methods

Research is the process of identifying a problem for study and reviewing the literature relating to the problem to develop a hypothesis or series of questions to guide examination of a problem under review. Once a hypothesis is developed the researcher establishes a method to interpret the collected data, which is evaluated and used by researchers to understand a phenomena.

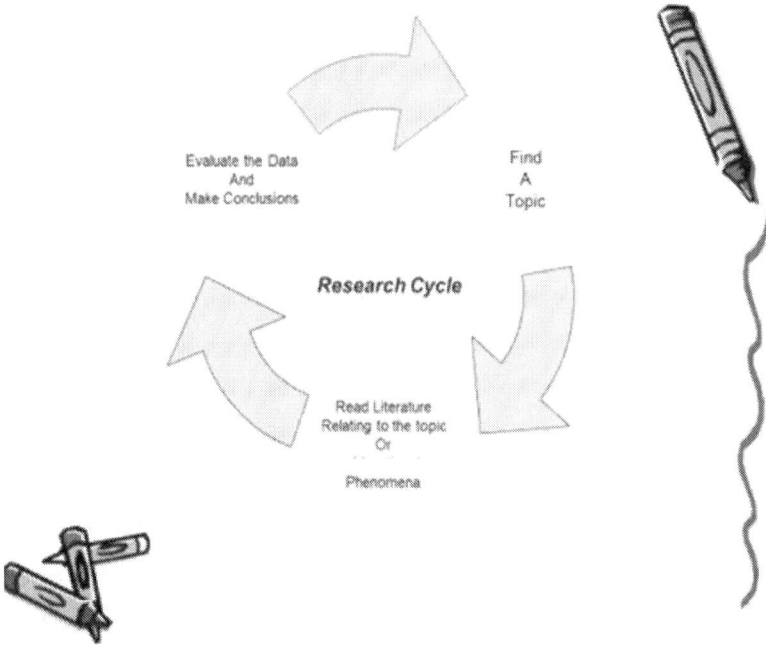

In population genetics the researcher usually uses the "wave of advance" model to explain demographic movements in the past. The "wave of advance" model was used to explain the spread of advantageous genes within a population (Ackland et al,2007; Renfrew, 2001).) . This theory was adapted to explain why an advantageous technology that may appear in one population spreads (and or taken)to another population living in a different geographical area (Ackland et al, 2007).

Ray and Excoffier (2009) and Renfrew (2001) argue that to build a reliable model of population dispersal researchers must combine genetic data and archaeological (or historical and linguistic) data .

The methods of Ray and Excoffier (2009) are in conformity with basic archaeogenetic research methods.

The Archaeogenetic method suggest that coupling the archaeological data with genetic data is a powerful way to infer population migration .

Although archaeogenetics was formerly the norm for many molecular geneticists during the first decade of the 21 Century, most researchers believe that Bayesian statistics alone, have sufficient power to demonstrate the valility of their research, and fail to corroborate the DNA data with corresponding archaeological, linguistic and paleo-anthropological evidence.

Failure to find corroborative evidence to support the genetic data makes claims regarding the ethnicity of ancient European populations suspicious. They are suspicious because the archaeological evidence does not support the claim of population geneticists that the Early Farmers (EF), and Steppe Pastoralists were Indo-Europeans

Many people don't know how to evaluate population genetics articles, because they are expost facto research based on " statistical inferences" or the beliefs of the researcher supported by statistics. As a result, researchers can not judge the validity and reliability of the research. One must assume the research is correct based solely on the Bayesian statistical inferences (Spiegelhalter and Rice, 2009)—not the interactions between an independent variable and dependent variable(s).

In research scientific there are two variables: one variables that can be manipulated and another variables that can not be manipulated. A variable that can be manipulated is a variable that can be changed for example, your ability to perform a particular task can be influenced by the amount of training you receive in performing the task.

A variable that can not be manipulated can not be changed . For example, right now you are a particular age, it can not be manipulated. You are either Black or white, race can not change.

Research studies include a number of variables. Variables can be manipulated or not manipulated. A Independent Variable (IV) is any variable used to control for individual differences (this variable usually not manipulated).

Dependent variable (DV) any outcome measure which is effected by the IV. The effect of sex (IV) on reading achievement (DV).

Validity is testing the appropriateness, meaningfulness and usefulness of specific inferences made from test scores. In qualitative research the extent to which the research uses methods and procedures that ensure a high degree of research quality and rigor.

Internal Validity, we assume that whatever was manipulated produced a change in the dependent measure. The IV insured by control of the extraneous variables: health, sex, race, SES, age, IQ, religion.

External validity, provides the ability to generalize the research findings. In other words the IV produced a change in the DV.

In normal scientific research the researcher states a hypothesis and uses the scientific method to test his/her hypothesis. The validity and reliability of the piece of research is then determined by statistical significance tests focused on the interaction between the independent and dependent variables.

In the traditional evaluation of a piece of research literature you look at the researcher's hypothesis, results and statistical methods s/he used to determine the statistical significance of the research. This is not the case in population genetics research; in this research you are evaluating statistical inferences based on ***the beliefs already held by the researcher*** about a set of data, instead of testing a hypothesis.

As a result, the research contained in a population genetics article, reflects the views and beliefs already held by the researcher. Thusly, the statistical inferences will automatically support the views and beliefs held by that researcher; and any outliners that fail to support the researcher's beliefs may not be mentioned in the resulting research article/paper.

Here we will ask the question: "How do you evaluate population genetics research if it is expost facto research, that lacks an experimental design?" First, we will attempt to look at the doxa that may influence a geneticist's research and the constructs that should be considered when evaluating this knowledge base.

In reading any piece of research literature, we assume that any article or book written by an establishment member of the academe is reliable and valid. A piece of research full of valid scientific and/or historical truths-- erudite scholarship and impeccable research based on the scientific method.

The *scientific method* is based on hypotheses testing. Hypotheses testing means that a researcher forms a hypothesis and test the hypothesis using a series of quantitative or qualitative statistical methods to determine the statistical significance of the hypothesis being tested. The scientific method is based on experimentation to test a hypothesis .

Population geneticists usually do not test hypotheses. They make inferences about data based on Bayesian statistical inferences. They do not use statistical methods to determine the statistical significance of a hypothesis, they use statistics to describe data being reviewed by the researcher based on the beliefs the researcher already holds about the data being reviewed.

Population genetics is a type of Expost facto research. Expost facto research design is a quasi-experimental type of study examining how an independent variable, present prior to the research study, affects a dependent variable.

Whereas the subjects in experimental research are randomly selected, the participants in Expost facto research , are not randomly selected or assigned.The genome of the research subjects is examined to determine the haplotypes and haplogroups carried by the participants in the study.

In population genetics research the researcher uses the Bayesian inference method of statistical inference. The Bayesian statistical method, is a subjective research design/method that provides a rational method of updating the researcher's beliefs.

Bayesian Methods

Bayesian statistics call for the researcher to identify a subjective prior distribution to begin their study. This makes Bayesian statistics a subjective research method. Geneticists can use Bayesian statistics to fake population movements because this method is used to make inferences about a data set. An inference is a conclusion reached on the basis of a researcher's evaluation and interpretation of the evidence. This is counter to the scientific method.

The scientific method is a method of inquiry based on empirical or measurable evidence gathered as a result of experimentation. This makes the scientific method objective, as opposed to the subjectivism of Bayesian statistics.

The scientific method is based on hypotheses testing. Hypotheses testing means that a researcher forms a hypothesis and test the hypothesis using a series of quantitative or qualitative statistical methods to determine the statistical significance of the hypothesis being tested. The scientific method is based on experimentation to test a hypothesis   .

Population geneticists usually do not test hypotheses. They base their research on prior distributions of their choice. Geneticists make inferences about data based on Bayesian statistical inferences.

They do not determine the statistical significance of a hypothesis to confirm their findings. Population geneticists use statistics to describe data being reviewed by the researcher based on the prior beliefs the researcher already holds about the data being reviewed.

The fact that the statistics are used to support the beliefs of the researcher, means that the results will confirm the opinion/inference/hypothesis the researcher already holds about the data. Thusly, there is no testing of a hypothesis, the research just confirms what they already believe to be true about a data set.

Population genetics is a type of Expost facto research. Expost facto research design is a quasi-experimental type of study examining how an independent variable, present prior to the research study, affects a dependent variable.

Whereas the subjects in experimental research are randomly selected, the participants in Expost facto research , are not randomly selected or assigned. The genome of the research subjects is examined to determine the haplotypes and haplogroups carried by the participants in the study.

In population genetics research the researcher uses the Bayesian inference method of statistical inference. Bayesian statistical method, is a subjective research design/method that provides a rational method of updating the researcher's beliefs.

Since, the results of a Bayesian statistical analysis are a series of beliefs based on statistical inferences, the results can not stand alone. This is due to the reality, that any results, reported by a researcher are only a series of inferences based on the researcher's belief about a phenomena backed up by a series statistical results. If the results are published without corresponding evidence from archaeology, anthropology, linguistics and or craniometrics the inferences are pure conjecture, because they reflect the attitudes already held by the researcher, confirmed by data selected by the researcher to support his or her beliefs.

This means that when a geneticists's provides their statistics supporting a line of reasoning, they are "proving" inferences they already hold about the data. This means that any data not supporting the inferences a researcher already made about the data, will be left out of the study.

This is one of the reasons some geneticists fail to support their Bayesian statistics with craniometric, linguistic and archaeological evidence. Failure to use this evidence to confirm the statistics, mean that you have to accept the data based upon hypothetical statistical measurements which may have little to do with natural conditions and phenomena. Thusly, you can fake the genetic history of a population and region by basing the history of a region solely on the haplogroups of the people who live in the area today, instead of the ancient DNA from ancient skeletons.

Since, the results of a Bayesian statistical analysis are a series of beliefs based on statistical inferences, the results can not stand alone. This is due to the reality, that any results, reported by a researcher are only a series of inferences based on the researcher's belief about a phenomena backed up by a series statistical results. If the results are published without corresponding evidence from archaeology, anthropology, linguistics and or craniometrics the inferences are pure conjecture, because they reflect the attitudes already held by the researcher, confirmed by data selected by the researcher to support his or her beliefs.

Doxa

There is a sociological basis behind how a researcher interprets data. Sociological research indicates that there are unconscious cognitive structures within each individual. Cognitive structures that hold the idealistic view of members of the academe that determine how they perceive "reality". These structures are called doxa (Berlinerblau 1999).

 Commenting on these schema Berlinerblau (1999) noted that "These types of theories share the assumption that human beings know things that they do not even know that they know; that they "possess" knowledge about the world which exists in some sort of cognitive substrate, beyond the realm of discourse" (p.106).Wacquant (1995) says that doxa is " a realm of implicit and unstated beliefs".

Given the research suggesting that doxa exist, support the view that some researchers allow their hatred of multiculturalism, ethnic prejudice and racism to define their discourse, teaching and writing about themes relating to groups " other" ,than their own cultural and ethnic group . Moreover, it suggest that when topics such as Eurasian and African haplogroups, Afrocentrism, African origins of the Dravidians and etc., is attacked by members of the academe, these academics are supported by the "establishment" without any reservation, or test of the validity of their claims.

In fact, it appears that doxic assumptions relating to the validity of back migration of so-called Eurasian genes into Africa, recent African origin of Dravidians and Dravidian origin of the Indus Valley Civilization obviates critique of the academics that disparage these themes. Due to Doxa you can state a researcher's attitude toward a historical, genetic or anthropological concept and theorems without the statement being an *ad hominem* .

Doxa exist in the study of Western European (WE) prehistory. Because Caucasians live in Western Europe. Geneticists believe that because Caucasians live in Western Europe today, they assume Caucasians have always been present in the region. The idea that "white" have always lived in Europe is Doxa.

Reference:

Ackland G J , Markus Signitzer, Kevin Stratford,and Morrel H. Cohen.(2007).Cultural hitchhiking on the wave of advance of beneficial technologies PNAS 104 (21) 8714-8719; published ahead of print May 16, 2007, doi:10.1073/pnas.0702469104. Retrieved 2/6/2015 at : http://www.pnas.org/content/104/21/8714.full

Berlinerblau, J. (1999). Heresy in the University: The Black Athena Controversy and the Responsibilities of American Intellectuals .Rutgers University Press.

Chaubey, G and Phillip Endicott. (2013)The Andaman Islanders in a Regional Genetic Context: Reexamining the Evidence for an Early Peopling of the Archipelago from South Asia. Retrieved 3/6/2015 at: http://digitalcommons.wayne.edu/cgi/viewcontent.cgi?article=2055&context=humbiol

Endicott P, Metspalu M, Stringer C, Macaulay V, Cooper A, et al. (2006) Multiplexed SNP Typing of Ancient DNA Clarifies the Origin of Andaman mtDNA Haplogroups amongst South Asian Tribal Populations. PLoS ONE 1(1):

e81.
http://journals.plos.org/plosone/article?id=10.1371/journ
al.pone.0000081

Kanakasabhai,V.(1966).   The Tamil Eighteen Hundred
Years ago .

Ray N, Excoffier L.2009. Inferring past demography using
spatially explicit population genetic models. Human
Biology, 81 (2-3): 141-157.

Reich, D. (2018). How Genetics Is Changing Our
Understanding of 'Race'. The New York Times,23 March .

Renfrew,   C.   (2001).From   molecular   genetics   to
archaeogenetics   PNAS   98   (9)   4830-4832;
doi:10.1073/pnas.091084198.
http://www.pnas.org/content/98/9/4830.full

Spiegelhalter, D. and Kenneth Rice. (2009). Scholarpedia,
4(8):5230.

Winters, C. (1985).   "The Far Eastern Origin of the
Dravidians", Journal of Tamil Studies, pp.66-92.

 Winters.C. (2010). Munda Speakers are the Oldest
Population in India. The Internet Journal of Biological
Anthropology.   2010   Volume   4   Number   2,
https://ispub.com/IJBA/4/2/5591

Who were the First Europeans

Based on Eurocentric Doxa that Caucasians have always lived in Europe has led to geneticists popularizing the idea that the Western Hunter Gatherers 15,000 years ago (15kya), Neolithic Early Farmers (EF) 7kya and Steppe Pastoralist (SP) 5yka, were ethnically caucasian or white.

The archaeological evidence makes it clear that the prehistoric Western Europeans were ethnically Negroes or Africans (Figure 1) .

In fact, the first anatomically modern European was Cro-Magnon man. Cro-Magnon man was "dark complexioned" and best represented by the Khoisan.

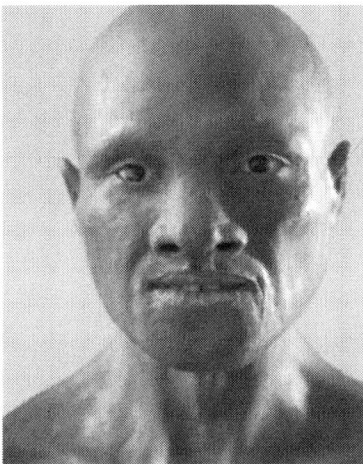

Figure 1. First European.

The archaeological research makes it clear Neanderthal probably mixed with Africans. As early as 200 kya homo sapien sapiens originated in Africa.

Probably 400 - 600 kya Homo rhodesiensis migrated into Europe. Homo rhodesiensis originated in Africa.

The ancestor of neanderthal man was Homo rhodesiensis. Between 139 kya and 125 kya the Neanderthals migrated back into Africa and spread from Morocco to East Africa (Ki-Zerbo, 1981: p. 572).

The Neanderthal used Mousterian tools. These tools were also being used in Africa as early 130 kya. This places Neanderthalers in North Africa.

The human types associated with the Neanderthal tools found at Jebel Ighoud and Haua Fteah resemble contemporaneous European Neanderthaler tools. The presence of Mousterian tools suggest that Neanderthalers mixed with Africans because we know that anatomically modern humans were living in the area at the time.

The African Neanderthal people used the common Levoiso-Mousterian tool kit originally discovered in Europe. The Nenderthal skeletons mainly come from Djebel Irhoud and El Guettar (Ki-Zerbo, 1981). Later Neanderthal people used the Aterian tool kit.

Pigmentation Gene

The popular view of the ancient Europeans is that they were pale-faced homo sapien sapiens. This is the view of Klyosov (2014).

Even though Klyokov (2014) maintains that the first anatomically modern human (AMH) were pale faced the research indicates a probable Paliolithic origin for the "light" pigment gene.

The functional alleles in the human pigmentation pathway include 3 genes: HERCZ, SLC24A5, and TYR. HERCZ provides the eye color to individuals. The HERCZ gene is found among dark skinned population like the Luyha and Maasai who live in East Africa.

Researchers have suggested that natural selection played a role in skin pigmentation. A popular hypothesis for light pigmentation in Europe is that the further away from the equator a population lives, due to decreasing exposure to sun light, in the high latitudes, decreased solar ultraviolet light results in decreased pigmentation. Decreased pigmentation allegedly results from greater synthesis of dermal vitamin D.

The pigmentation center is SLC24A5. The ancestral gene for light skin rs1426L54 is "predominant" among sub-Saharan African (SSA) populations (Canfield et al., 2014). The derived allele from this coding polymorphism for light skin is A111T alleles (Canfield et al., 2014). The A111T pigmentation haplotype indicate high frequencies among "light skinned" populations in Europe and East Asia.

The highest frequencies of SLC24A5A111T are found in Western Europeans. Interestingly, the lowest global frequencies of SLC24A5A111T are found in SSA, the Polar Region and East Asia (EA) (Canfield et al., 2014).

The frequency of A111T haplotype in EA and SSA is 0.15, while in Western Eurasia the average frequency of this mutation is 0.9 - 1.0 (Canfirld et al., 2014). The low frequency of A111T among non-western Eurasians suggest that this mutation may only account for pale skin pigmentation in Western Eurasia.

Origin of Dark Skinned AMH via IBERIA or the Levant

The evidence makes it clear that the first Europeans were dark skinned. Olalde et al. (2014) provides conclusive genetic evidence that hunter gatherers in Mesolithic Europe were dark skinned or highly pigmented with light (blue) eyes. The Olalde et al. (2014) study of La Brana-Arintero site in Leon, Spain of dark skinned hunter-gatherer Europeans corresponds to the Loschbour sample from Luxembourg, of dark skinned Europeans.

Cassidy (2020) reported that the Irish dating to around 4-4700kya were also dark skinned along with other European population as illustrated in Table 12 of this article. This cline of pigmentation in western Eurasia appears to be associated with Cro-Magnon man, the first AMH in western Europe who was associated with the Aurignacian culture (see Figure 1).

It is not uncommon for Blacks to be dark skinned with blue eyes and blond hair (see Figure 2). This is especially true for Blacks in Melanesia.

Dark Skinned People with Blue Eyes and Blond Hair

Figure 2

The first AMH European reconstructed by Forensic artist Richard Neave, of National Geographic from 35kya resembled a Khoisan individual (see Figure 1). This supports the research of Boule and Vallois (1957) that South Africans migrated across Africa, into Europe over 35 kya.

Most researchers maintain that the first AMH European came from the Levant. This migration of AMH entering western Eurasia from the East is not supported by the archaeological evidence. The Auriganacian culture which is associated with the Cro-Magnon people crossed the Straits of Gibraltar (SG) from Africa into Iberia (Winters, 2008,

2011).

We know that the first AMH probably entered western Eurasia via the SG, because Neanderthals dominated the Levant until around 30 - 20 kya. Between 10 - 5 kya the Levant population was "tropically adapted" hominids, especially in relation to Qafzed-Skhul (QS) hominids (Holliday, 2000). Ninety-five percent of the QS population were SSA (e.g., Qafzeh 8 at 85%, and Skhul 4 at 71%) (Holliday, 2000).

The fauna and zooarchaeological remains from QS, indicate the hominids here exploited African fauna (Holliday, 2000). Holliday (2000) claims the QS people were Proto-Cro-Magnons, because they were similar in dental and craniological size to the Aurignacian hominids (Holliday, 2000). Except of the AMH at QS, the majority of hominids in the Levant were Neanderthal.

The craniometric evidence for SSA populations in Europe was also examined by Brace et al. (2006). Brace et al. (2006) after studying 24 craniofacial measurements of AMH from Europe was surprised to find that Neolithic people in Europe fail to be related to modern Europeans. Some researchers have assumed that the Basque, a non-Indo-European population in Spain probably represented descendants of the original Europeans, but samples from this group and Canary Islanders did not correspond to the Natufians or Cro-Magnon populations (Brace, 2006).

The founders of civilization in Southwest Asia were the people, archaeologists call Natufians. By 13,000 BC, according to Clark (1977) the Natufians were collecting grasses in Nubia (Ehret, 1979), which later became domesticated crops in Southwest Asia. In Palestine the Natufians established intensive grass collection. The Natufians used the Ibero-Maurusian tool industry (Wendorf, 1968).

The Aurignacian civilization was founded by the Cro-Magnon people who originated in Iberia. They took this culture to Western Europe across the Straits of Gibraltar (Winters, 2011). The Cro-Magnon people were probably Bushman/ Khoisan (Boule & Vallois, 1957).

The "Classic Aurignacian" culture probably began in Africa, crossed the Straits of Gibraltar into Iberia, and expanded eastward across Europe (Brown, 2006; Gilead, 2005; Steven et al., 2001; Winters, 2008, 2011). The archaeological record informs us that Cro-Magnon people replaced the Neanderthal population of the Levant, at Ksar Akil around 32,000 years ago (Steven et al., 2001; Gilead, 2005), not the Natufians who entered the Levant almost 20,000 years later. This indicates that the Aurignacians moved west to east from Iberia across Europe.

The archaeological and craniometric evidence indicates that the pre-Indo-European people were probably highly pigmented. There have been numerous "Negroid skeletons" found in Europe according to Boule and Vallois (1957). Diop (1991) discussed the Negroes of Europe in Civilization or Barbarism (pp. 25-68). Also W.E. B. DuBois, The World and Africa noted that "There was once a an "uninterrupted belt' of Negro culture from Central Europe to South Africa" (p. 88).

Boule and Vallois (1957) reported the find of SSA skeletons at, Grotte des Enfants, Chamblandes in

Switzerland, several Ligurian and Lombard tombs of the Metal Ages have also yielded evidences of a Negroid element.

Since the publication of Verneau's memoir, discoveries of other Negroid skeletons in Neolithic levels in Illyria and the Balkans have been announced. The prehistoric statues, dating from the Copper Age, from Sultan Selo in Bulgaria are also thought to protray Negroids (Boule & Vallois, 1957).

In 1928 Rene Bailly found in one of the caverns of Moniat, near Dinant in Belgium, a human skeleton of whose age it is difficult to be certain, but seems definitely prehistoric. It is remarkable for its Negroid characters, which give it a reseblance to the skeletons from both Grimaldi and Asselar (Diop, 1991).

Boule and Vallois (1957), was able to chart the migration of civilization from South Africa to the Aurignacian culture of Europe. These anthropologist reported that the Khoisan shared the same style stone implements and burials "associated with the Aurignacian or Solutrean type industry..." (Boule & Vallois, 1957: pp. 318-319).

They add, that in relation to Bushman [Khoisan] art "This almost uninterrupted series leads us to regard the African continent as a centre of important migrations which at certain times may have played a great part in the stocking of Southern Europe. Finally, we must not forget that the Grimaldi Negroid skeletons show many points of resemblance with the Bushman [Khoisan] skeletons" (Boule & Vallois, 1957).

There were numerous out of Africa exits into Iberia. The African migrants introduced into Europe, the Aurignacian, Solutrean , Bell Beaker/Corded ware and Moorish cultures between 44,000 BC and 1492 AD. These Sub-Saharan populations belonged to the Black Variety(1) .

The African population includes a variety of African/Black populations that range from the Khoisan and Pygmy hunter-gatherer populations , to Blacks with both narrow and broad features. The variety of African populations has resulted in the existence of a variety of genomes carried by these populations since the origination of anatomically modern humans in Africa over the past 200,000 years (200ky). Today the Black Variety is referred to in the archaeogenetics literature as Sub-Saharan Africans (1).

Between 3200-2900 BC, African culture and people began to migrate into Iberia and introduced megaliths and the Bell Beaker culture (1,9). Spanish researchers accepted the reality that the Iberia Peninsula owed the major parts of Neolithic Iberia to African immigrants (1-3).

The recent article by Fregel et al (4) noted that all the burials at the Ifri n'Amr o'Moussa site IAM1-IAM7 , are devoid of any artifacts, except for an original funeral ritual, which consists of placing a millstone on the skull . These burials were dated from 4,850 to 5,250 BCE, they carried U6, M1, T2, X and K . This suggest that Africans were already carrying this mtDNA(4) .

MacWhite (3) claims there was a close relationship between Iberia and Britain. These researchers admit that Portugal and Brittany were settled by Megalithic Africans who founded respectively the Mugem and Teviec sepultures (2-

3). The Chalcolithic Iberian civilizations of Los Millares and Zambujal is dated to 2600BC. The artifacts from these cultures correspond to the North African sites of Kehf el Baroud site ((c. 2,800 ± 110- 3,210 ± 110), the necropolis of Rouazi-Skhirat and El Kiffen (5). The white layer at these sites corresponds to a similar layer at Los Millares and Vila Nova de Sao Pedro (5).

Bell Beaker appeared in Iberia around 2700 BCE (6-9). It is interesting to note that while most people in the Iberian Beaker complex carried the G2 and I2a2 haplogroups (8,10). Iberians during this period also carried R-V88 (10) . The Bell Beaker cultures in Iberia , especially in the El Toro cave was occupied only occasionally, particularly at the end of the 3rd and 2nd millennium BC . The Iberian Late Copper Age, with Bell Beakers, and the Bronze Age artifacts dates back to this period.

Iñigo Olalde et al (7) discuss the spread of Bell Beaker culture across Europe 2.7 kya. These researchers found limited genetic affinity between individuals from Iberian and central Europeans. Iñigo Olalde et al (7) concludes that migration probably played an insignificant mechanism in the spread of R1 within the two areas.

The Neolithic British farmers were genetically similar to Neolithic Iberians dating between 3900–1200 BCE (1-2, 9-10) . The British farmers were replaced by farmers of the Beaker culture (7). Eighty-four percent of the Beaker Bell Steppe migrants carried R1b (7).

In summary Late Neolithic Bell Beaker tradition originated in Africa and expanded from the Taqua region of These Iberia to Ireland and Scandinavia between 2800-2700 BC.

The Y-Chromosome carried by these people was primarily R1b1 (R-L278) at Samara and in Spain.

Haplogroups in Ancient Europe

Most of the haplogroups associated with Eurasian populations are derived from L3(M, N). Winters (2010) has argued that L3(M, N) spread across Africa before the Out of Africa (OoA) event 60 kya.

Researchers have found that western Eurasians carry some Neanderthal genes. There is little evidence of Neanderthal genes among African populations. An exception to this norm are the Khoisan who share a phylogenic relationship with Altai Neanderthals (Prufer et al., 2014).

The traditional view for the spread of L3(M, N) across Eurasia is that the M and N macrohaplogroups originated in western Eurasia and returned to Africa as a result of back-migration. The big problem for this theory is that the proposed dates for the origin of haplogroups N and M in western Europe, date to a period when these areas were inhabited by Neanderthal people—not AMH. This supports an African origin for L3(M, N).

The craniometric evidence supports a Khoisan presence in Europe during Aurignacian times. If the Khoisan represent the ancient dark skinned European population, this reality should be able to be confirmed by genetic research.

The most archaic AMH remains come from Florished, South Africa; they date between 190 - 330 kya (Rito et al., 2013). Other ancient fossil evidence of AMH in South Africa come from Broken Hill (c. 110 kya) and the Klasis

River caves (c. 65 - 105 kya).

Researchers have been surprised to find Khoisan and European admixture. The idea that the Khoisan acquired Eurasian admixture via Ethiosemitic speakers is pure speculation (Pickrell et al., 2013). There is no archaeological evidence of Ethiopians migrating into East and South Africa, but there is evidence of an ancient migration of Khoisan into Europe based on archaeological and skeletal data.

The Khoisan carry haplogroups L3(M, N). Before they reached Iberia, they probably stopped in West Africa. The basal L3(M) motiff in West Africa is characterized by the Ddel site np 10,394 and Alul site np 10,397 associated with AF-24. This supports my contention that Khoisan speakers early settled West Africa on their way to Iberia.

Granted L3 and L2 are not as old as LOd, but Gonder et al. (2006) provides very early dates for this mtDNA e.g., L3(M, N) 94.3; the South African Khoisan (SAK) carry L1c, L1, L2, L3 M, N dates to 142.3 kya; the Hadza are L2a, L2, L3, M, N, dates to 96.7 kya.

The dates for L1, L2, L3, M, N are old enough for the Khoisan to have taken N to West Africa, where we find L3, L2 and LOd and thence to Iberia as I suggested in my paper (Winters, 2011).

It is interesting to note that LO haplogroups are primarily found among Khoisan and West Africans. This shows that at some point in prehistory the Khoisan had migrated into West Africa.

The first modern European reconstructed by Forensic artist Richard Neave based on skull fragments from 35,000 years ago resembled a Khoisan (Figure 1). The skull was discovered in the southwest region of Romania's Carpathian Mountains. This supports the research of Boule and Vallois that South Africans migrated into Europe 35 kya. This genetic evidence now supports Boule and Vallois of a Khoisan migration into Europe.

The Khoisan may have introduced the L haplogroup to Iberia. The SAK populations carry haplogroups L2, and L3. de Domínguez (2005), noted that much of the ancient mtDNA found in Iberia has no relationship to the people presently living in Iberia today and correspond to African mtDNA haplogroups. de Domínguez (2005) found that the lineages recovered from ancient Iberian skeletons are the African lineages L1b, L2 and L3. Almost 50% of the lineages from the Abauntz Chalcolithic deposits and Tres Montes, in in Bulgaria are also thought to protray Negroids (Boule & Vallois, 1957).

In 1928 Rene Bailly found in one of the caverns of Moniat, near Dinant in Belgium, a human skeleton of whose age it is difficult to be certain, but seems definitely prehistoric. It is remarkable for its Negroid characters, which give it a resemblance to the skeletons from both Grimaldi and Asselar (Diop, 1991).

Boule and Vallois (1957), was able to chart the migration of civilization from South Africa to the Aurignacian culture of Europe. These anthropologist reported that the Khoisan shared the same style stone implements and burials "associated with the Aurignacian or Solutrean type industry..." (Boule & Vallois, 1957: pp. 318-319).

They add, that in relation to Bushman [Khoisan] art "This almost uninterrupted series leads us to regard the African continent as a centre of important migrations which at certain times may have played a great part in the stocking of Southern Europe. Finally, we must not forget that the Grimaldi Negroid skeletons show many points of resemblance with the Bushman [Khoisan] skeletons" (Boule & Vallois, 1957).

Most of the haplogroups associated with Eurasian populations are derived from L3(M, N). Winters (2010) has argued that L3(M, N) spread across Africa before the Out of Africa (OoA) event 60 kya.

Researchers have found that western Eurasians carry some Neanderthal genes. There is little evidence of Neanderthal genes among African populations. An exception to this norm are the Khoisan who share a phylogenic relationship with Altai Neanderthals (Prufer et al., 2014).

The traditional view for the spread of L3(M, N) across Eurasia is that the M and N macrohaplogroups originated in western Eurasia and returned to Africa as a result of back-migration. The big problem for this theory is that the proposed dates for the origin of haplogroups N and M in western Europe, date to a period when these areas were inhabited by Neanderthal people—not AMH. This supports an African origin for L3(M, N).

The craniometric evidence supports a Khoisan presence in Europe during Aurignacian times. If the Khoisan represent the ancient dark skinned European population, this reality should be able to be confirmed by genetic research.

The most archaic AMH remains come from Florished, South Africa; they date between 190 - 330 kya (Rito et al., 2013). Other ancient fossil evidence of AMH in South Africa come from Broken Hill (c. 110 kya) and the Klasis River caves (c. 65 - 105 kya).

Researchers have been surprised to find Khoisan and European admixture. The idea that the Khoisan acquired Eurasian admixture via Ethiosemitic speakers is pure speculation (Pickrell et al., 2013). There is no archaeological evidence of Ethiopians migrating into East and South Africa, but there is evidence of an ancient migration of Khoisan into Europe based on archaeological and skeletal data.

The Khoisan carry haplogroups L3(M, N). Before they reached Iberia, they probably stopped in West Africa. The basal L3(M) motiff in West Africa is characterized by the Ddel site np 10,394 and Alul site np 10,397 associated with AF-24. This supports my contention that Khoisan speakers early settled West Africa on their way to Iberia.

Granted L3 and L2 are not as old as LOd, but Gonder et al. (2006) provides very early dates for this mtDNA e.g., L3(M, N) 94.3; the South African Khoisan (SAK) carry L1c, L1, L2, L3 M, N dates to 142.3 kya; the Hadza are L2a, L2, L3, M, N, dates to 96.7 kya.

The dates for L1, L2, L3, M, N are old enough for the Khoisan to have taken N to West Africa, where we find L3, L2 and LOd and thence to Iberia as I suggested in my paper (Winters, 2011).

It is interesting to note that LO haplogroups are primarily found among Khoisan and West Africans. This shows that at some point in prehistory the Khoisan had migrated into West Africa.

The first modern European reconstructed by Forensic artist Richard Neave based on skull fragments from 35,000 years ago resembled a Khoisan (Figure 1). The skull was discovered in the southwest region of Romania's Carpathian Mountains. This supports the research of Boule and Vallois that South Africans migrated into Europe 35 kya. This genetic evidence now supports Boule and Vallois of a Khoisan migration into Europe.

The Khoisan may have introduced the L haplogroup to Iberia. The SAK populations carry haplogroups L2, and L3. de Domínguez (2005), noted that much of the ancient mtDNA found in Iberia has no relationship to the people presently living in Iberia today and correspond to African mtDNA haplogroups.

de Domínguez (2005) found that the lineages recovered from ancient Iberian skeletons are the African lineages L1b, L2 and L3. Almost 50% of the lineages from the Abauntz Chalcolithic deposits and Tres Montes, in In Europe only 0.2 of the population belong to haplogroup N. The carriers of haplogroup N are mainly situated in Central Europe.

In Africa the N haplogroup is found throughout the African continent. Sub-Saharan African populations carrying haplogroup N belong to almost all the language families spoken in Africa including Cushitic, Nilo-Saharan, Khoisan, Niger-Congo, and Semitic.

Stefania et al (2019) noted that "Here we present newly obtained mitochondrial genomes from two ~7000-yearold individuals from Takarkori rockshelter, Libya, representing the earliest and first genetic data for the Sahara region. These individuals carry a novel mutation motif linked to the haplogroup N root. Our result demonstrates the presence of an ancestral lineage of the N haplogroup in the Holocene "Green Sahara", associated to a Middle Pastoral (Neolithic) context". This supports the view that the macrohaplogroup N, originated in Africa.

The majority of carriers of haplogroup N in Africa live in Sub-Saharan Africa. In East Africa we find 85.5 percent of the populations carrying haplogroup N. Another 14.5 percent of the carriers of haplogroup N live in West Africa. The contemporary genomic data for haplogroup N in Africa and archaeological data indicates that this haplogroup probably appeared first in East Africa near the Great Lakes region. The geographical center for haplogroup N was probably Tanzania. Here we find in relative close proximity speakers of Khoisan, Niger-Congo, Cushitic and Nilo-Saharan language families that carry haplogroup N (Winters, 2010).

The Khoisan carry haplogroups L3(M, N). Before they reached Iberia, they probably stopped in West Africa.
The basal L3(M) motiff in West Africa is characterized by the Ddel site np 10394 and Alul site np 10397 associated with AF-24. This supports the view that Khoisan speakers early settled West Africa on their way to Iberia.

Granted L3 and L2 are not as old as LOd, but Gonder et al. (2006) provides very early dates for this mtDNA e.g.,

L3(M, N) 94.3; the SAK carry haplogroups L1c, L1, L2, L3 M, N and dates to 142.3 kya; the Hadza are L2a, L2, L3, M, N, and dates to 96.7 kya.

The dates for L1, L2, L3, M, N are old enough for the Khoisan to have taken N to West Africa and thence Iberia.This genetic evidence now supports Boule and Vallois (1957) of a khoisan migration into Europe.

The Y-chromosome denotes the male haplogroups. Y-chromosome haplogroup A is represented among African populations. Haplogroup A has its highest frequencies among the Khoisan and Pgymies. In Table 1, we present the percentage of South African Khoisan who carry haplogroups A and B. Haplogroup A is around 140,000kya.

In West Africa, under 5% of the NC speakers belong to group A. Most Niger-Congo speakers who belong to group A are found in East Africa and belong to A3b2-M13: Kenya (13.8) and Tanzanian (7.0%).

The second oldest Y-chromosome is haplogroup B. Haplogroup B is common among the forest people: the Pygmy groups namely the Baka and Mbuti.

Haplogroup A is largely restricted to parts of Africa, though a handful of cases have been reported in Europe and Western Asia. The clade achieves its highest modern frequencies in the Bushmen hunter-gatherer populations of Southern Africa, followed closely by many Nilotic groups in Eastern Africa. However, haplogroup A's oldest sub-clades are exclusively found in Central-Northwest Africa, where it, and consequently Y-chromosomal Adam, is believed to have originated about 140,000 years ago. The clade has also been observed at notable frequencies in certain populations in Ethiopia, as well as some Pygmy groups in Central Africa.

In a composite sample of 3551 African men, Haplogroup A had a frequency of 5.4%. The highest frequencies of haplogroup A have been reported among the Khoisan of Southern Africa, Beta Israel, and Nilo-Saharans from Sudan.

Haplogroup A has been observed as A1 in European men in England. As A3b2, it has been observed with low frequency in Asia Minor, the Middle East, and some Mediterranean islands, among Aegean Turks, Sardinians, Palestinians, Jordanians, Yemenites, and Omanis. Without testing for any subclade, haplogroup A has been observed in a sample of Greeks from Mitilini on the Aegean island of Lesvos and in samples of Portuguese from southern Portugal, central Portugal, and Madeira.

Table 1. Khoisan y-chromosome haplogroups[*].

Norm	A2 haplogroup	A3b1 haplogroup	B26 haplogroup
64 kung	8%	18%	8%
26 khwe		12%	
7 san	43%		57%

[*]After Scozzari et al., 2014.

Haplogroup B (YDNA) is localized to sub-Saharan Africa,

especially to tropical forests of West-Central Africa. After Y-haplogroup A, it is the second oldest and one of the most diverse human Y-haplogroups. It was the ancestral haplogroup of not only modern Pygmies like the Baka and Mbuti, but also Hadzabe from Tanzania, who often have been considered, in large part because of some typological features of their language, to be a remnant of Khoisan people in East Africa.

Controversy surrounds the migration of AMH into western Eurasia. Olalde et al. (2014) believes that the La Bana samples indicates "the existence of a common ancient genome signature across western and central Eurasia from the Upper Paleolithic to the Mesolithic.

The La Bana population belonged to y-chromosome C6. The y-haplogroup C6 finding in Mesolithic Iberia supports the early introduction of AMH across the Straits of Gibraltar—not eastern Eurasia.

The Y-DNA of the Irish population was haplogroup H, which is spread from South Asia to Ireland (Cassidy et al,2020). Many ancient Europeans from the Mesolithic to the Bronze Age carried Y-DNA I2 (Cassidy et al,2020).

The ancestral alleles from La Bana and Luxemburg dark skinned Europeans (Olalde et al., 2014), and the dark skinned Irish people (Cassidy et al, 2020) make it clear early Europeans were not pale skinned as Klyosov (2014) alleges. This genetic evidence for dark pigmented ancient Europeans was supported by the negro skeletons associated with ancient European sites (Boule & Vallois, 1957).

Discussion

Archaeological evidence from Africa details the expansion of AMH across Africa. There was probably a serial expansion of haplogroup N across Africa into Eurasia (Winters, 2010). Haplogroup N probably originated in the Great Lakes region of East Africa 93.4 kya (Winters, 2010). From Tanzania, Khoisan speaking people probably spread the haplogroup into Ethiopia by 80 kya. West Africa at this time and the Sahara was much wetter. This suggest that there may have been considerable threat of diseases such as sleeping sickness and sickle cell anemia; and as a result these areas were sparsely populated and haplogroup N did not spread into these areas until 70 kya (Winters, 2010).

Due to population increases in Ethiopia and other parts of east Africa 60 kya Sub-Saharan Africans carrying haplogroup N migrated into Yemen and on into East Eurasia (Winters, 2010).

The craniofacial evidence makes it clear that the Levantines and Ancient Europeans came from Africa (Brace et al., 2006; Holliday, 2000; Winters, 2011). They introduced SSA flora and fauna into Eurasia (Holliday, 2000).

As a result we find those craniofacial features of the Grimaldi-Cro-Magnon population (Brace et al., 2006; Barral & Charles, 1963), were shared with the Natufian population when plotted, and fall within the range of Sub-Saharan populations like the Niger-Congo speakers (Balter, 2005).

Numerous Sub-Saharan skeletons have been found in Europe dating to the Aurignacian and Neolithic periods (Brace et al., 2006; Boule & Vallois, 1957; Diop, 1974, 1991; DuBois, 1941). Boule and Vallois (1957) observed that Sub-Saharan skeletons have been found in the Ligurian and Lombard tombs, Grotte des Enfants, Chamblandes in Switzerland, caverns of Moniat, near Dinant in Belgium.

 Boule and Vallois (1957) claim that these European farmers correspond to the Khoisan population. This is interesting because Brace et al. (2006) found that the craniofacial features of these early European farmers and the Natufians plotted with Sub-Saharan groups (Brace et al., 2006) just like the Aurignacians (Boule & Vallois, 1957; Winters, 2011).

Skoglund et al. (2014) investigated the pigmentation of ancient Europeans including skeletal remains from Ajvide 5, La Brana 1, and the Iceman. The analysis by Skoglund et al. (2014) determined that the pigmentation phenotype for these Europeans was dark skin.

There are N hgs found in Africa. Haplogroups N, N* and N1 is found in low frequencies within Sub-Saharan groups including Senegambians (Gonzalez et al., 2006), Tanzanians (Gonder et al., 2006) and modern Ethiopians (Quibtanana-Murci, 1999). In Egypt 8.8 percent of the Gurma carry hg N1b (Stevanovitch et al., 2003). Stefania et al (2019) reported the discovery of a 70,000 Year Old Ancestral mitochondrial N lineage from the Neolithic 'green' Sahara.

Conclusion

In conclusion, the ancient Europeans were dark skinned (Lazaridis et al., 2013; Olalde et al., 2014). They carried mtDNA haplogroups H, N, and U, and probably y-chromosomes A and C6. Some of these Blacks had blue eyes (Lazaridis et al., 2013).

Neanderthals lived in Africa at Jebel Ighoud and Haua Fteah (Ki-Zerbo, 1981). The Khoisan carried Neanderthal genes (Scozzari et al., 2014).

These Black Europeans carried haplogroups H and N. These haplogroups continue to be carried by Sub Saharan Africans (Winters, 2010). This is based on the reality that the haplogroup N1(a) is common to Senegambians, modern Ethiopians and the Dravidian speaking people of India; and the craniometric evidence indicated that the Aurignacian and Neolithic populations were Sub-Saharan Africans (Boule & Vallois, 1957; Diop, 1991).

Thus, the ancient hunter-gather Europeans and European farmers were related to African groups. These dark skinned people probably planted the seeds of agriculture in ancient Europe. Interestingly, between 23,000-7000 BC the dominant haplogroup of Western Eurasians remained hg N1 (Winters, 2011).

References

Balter, M. (2005). Ancient DNA Yields Clues to the Puzzle of European Origins. Science, 310, 964-965. http://dx.doi.org/10.1126/science.310.5750.964

Barral, L., & Charles, R. P. (1963). Nouvelles donnees anthropometriques et precision sue les affinities systematiques des negroides de Grimaldi. Bulletin du Musee d'Anthropologie Prehistorique de Monaco, 10, 123-139.

Boule, M., & Vallois, H. V. (1957). Fossil Man. New York: Dryden Press.

Brace, C. L., Seguchi, N., Quintyn, C. B., Fox, S. C., Nelson, A. R., Manolis, S. K., & Pan, Q. F. (2006). The Questionable Contribution of the Neolithic and the Bronze Age to European Craniofacial Form. Proceedings of the National Academy of Sciences of the United States of America, 103, 242-247. http://dx.doi.org/10.1073/pnas.0509801102

Brown, S. J. (2006). Neanderthals and Modern Humans in Western Asia. http://karmak.org/archive/2003/01/westasia.html

Canfield, V. A., Berg, A., Peckins, S. et al. (2014). Molecular Phylogeography of a Human Autosomal Skin Color Locus under Natural Selection. G3, 3, 2059-2067. http://dx.doi.org/10.1534/g3.113.007484

Caramelli, D., Lalueza-Fox, C., Vernesi, C., Lari, M., Casoli, A., Mallegni, B. C., Dupanloup, I., Bertranpetit, J., Barbujani, G., & Bertorelle, G. (2003). Evidence for a Genetic Discontinuity between Neandertals and 24,000 Year-Old Anatomically Modern Europeans. Proceedings of the National Academy of Sciences of the United States of America, 100, 6593-6597. http://dx.doi.org/10.1073/pnas.1130343100

Caramelli, D., Milani, L., Vai, S., Modi, A., Pecchioli, E. et al. (2008). A 28,000 Years Old Cro-Magnon mtDNA Sequence Differs from All Potentially Contaminating Modern Sequences. PLoS ONE, 3, e2700. http://dx.doi.org/10.1371/journal.pone.0002700

Cassidy, L.M., Maoldúin, R.Ó., Kador, T. et al. (2020). A dynastic elite in monumental Neolithic society. Nature 582, 384–388 (2020). https://doi.org/10.1038/s41586-020-2378-6

Clark, J. D. (1977). The Origins of Domestication in Ethiopia. 5th Panafrican Congress of Prehistory and Quaternary Studies, Nairobi. de Domínguez, E. F. (2005). Polimorfismos de DNA mitocondrial en poblaciones antiguas de la cuenca mediterránea. PhD

Thesis, Barcelona: Universitat de Barcelona, Departament Biologia Animal.

Diop, A. (1974). The African Origin of Civilization. Brooklyn, NY: Lawrence Hill Books.

Diop, A. (1991). Civilization or Barbarism. Brooklyn, NY: Lawrence Hill Books.

DuBois, W. E. B. (1941). The World and Africa.

Ehret, C. (1979). On the Antiquity of Agriculture in Ethiopia. Journal of African History, 20, 161-177. http://dx.doi.org/10.1017/S002185370001700X

Gilead, I. (2005). The Upper Paleolithic Period in the Levant. Journal of World Prehistory, 5, 105-154.

Gonder, M. K., Mortensen, H. M., Reed, F. A., de Sousa, A., & Tishkoff, S. A. (2006). Whole mtDNA Genome Sequence Analysis of Ancient African Lineages. Molecular Biology and Evolution, 24, 757-768. http://dx.doi.org/10.1093/molbev/msl209

González, A. M., Cabrera, V. M., Larruga, J. M., Tounkara, A., Noumsi, G., Thomas, B. N., & Moulds, J. M. (2006). Mitochondrial DNA Variation in Mauritania and Mali and Their Genetic Relationship to Other Western Africa Populations. Annals of Human Genetics, 70, 631-657. http://www.blackwell-synergy.com/doi/abs/10.1111/j.1469-

1809.2006.00259.x?cookieSet=1&journalCode=ahg
http://dx.doi.org/10.1111/j.1469-1809.2006.00259.x

Haak, W., Forster, P., Bramanti, B., Matsumura, S., Brandt, G., Tänzer, M., Villems, R., Renfrew, C., Gronenborn, D., Alt, K. W., & Burger, J. (2005). Ancient DNA from the First European Farmers 7500-Year-Old Neolithic Sites. Science, 310,1016-1018.

Holliday, T. (2000). Evolution at the Crossroads: Modern Human Emergence in Western Asia. American Anthropologist, 102, 54-68.

Ki-Zerbo, J. (1981). Unesco General History of Africa Vol. 1: Methodology and African Prehistory. 572.

Klyosov, A. A. (2014). Reconsideration of the "Out of Africa" Concept as Not Having Enough Proof. Advances in Anthropology, 4, 18-37.
http://www.scirp.org/journal/aa
http://dx.doi.org/10.4236/aa.2014.41004

Lazaridis, J., Patterson, N., Mittnik, A. et al. (2013). Ancient Human Genomes Suggest Three Ancestral Populations for Present-Day Europeans.
http://biorxiv.org/content/biorxiv/early/2013/12/23/00 1552.full.pdf

Olalde, I., Allentoft, M. E., Sanchez-Quinto, F., Santpere, G., Chiang, C. W. K., DeGiorgio, M. et al. (2014). Derived Immune and Ancestral Pigmentation Alleles in a 7,000-Year-Old Mesolithic European. Nature, 507, 225-228.
http://dx.doi.org/10.1038/nature12960

Pickrell, J. K., Patterson, N., Loh, P. R., Lipson, M., Berger, B., Stoneking, M., Pakendorf, B., & Reich, D. (2013). Ancient West Eurasian Ancestry in Southern and Eastern Africa. http://arxiv.org/abs/1307.8014

Prufer, K., Racimo, F., Patterson, N., Jay, F., Sankararaman, S., Sawyer, S. et al. (2014). The Complete

Genome Sequences of Neanderthal from the Altai, Mountains. Nature, 505, 43-49.

Quibtanana-Murci, L., Semino, O., Bandelt, H. J., Passaro, G., McElreadey, K., & Santachiara-Benerecetti, A. S. (1999). Genetic Evidence of an Early Exit of Homo Sapiens from Africa through Eastern Africa. Nature Genetics, 23, 437-441. http://dx.doi.org/10.1038/70550

Rito, T., Richard, M. B., Fernandes, V., Alshamal, F., Cerny, V., Pereira, L., & Soares, P. (2013). The First Modern Human Dispersals aross Africa. PLoS ONE, 8, e80031.

Scozzari, R., Massaia, A., Trombatta, B., Bellusci, G., Myres, N. M., Novelletto, A., & Cruciani, F. (2014). An Unbiased Resource of Novel SNP Markers Provides a New Chronology for Human Y-Chromosome and Reveals a Deep Phylogenetic Structure in Africa. Genome Research.

Skoglund, P., Malmström, H., Omrak, A., Raghavan, M., Valdiosera, C., Günther, T., Hall, P., Tambets, K., Parik, J., Sjögren, K. G., Apel, J., Willerslev, E., Storå, J., Götherström, A., & Jakobsson, M. (2014). Genomic Diversity and Admixture Differs for Stone-Age Scandinavian Foragers and Farmers. Science, 344, 747-750. http://dx.doi.org/10.1126/science.1253448

Stefania Vai , Stefania Sarno, Martina Lari , Donata Luiselli3, Giorgio Manzi, Marina Gallinaro , Safaa Mataich, Alexander Hübner, Alessandra Modi1, Elena Pilli, MaryAnneTafuri, David Caramelli & Savino di Lernia .(2019). Ancestral mitochondrial N lineage from the Neolithic 'green' Sahara . https://www.nature.com/articles/s41598-019-39802-1.pdf

Stevanovitch, A., Gilles, A., Bouzaid, E., Kefi, R., Paris, F., Gayraud, R. P., Spadoni, J. L., El-Chenawi, F., & Béraud-Colomb, E. (2003). Mitochondrial DNA Sequence Diversity in a Sedentary Population from Egypt. Annals of

Human Genetics, 68, 23-29. http://dx.doi.org/10.1046/j.1529-8817.2003.00057.x

Steven, L. K., Stiner, M. C., Reese, D. S., & Gulec, E. (2001). Ornaments of the Earliest Upper Paleolithic: New Insights from the Levant. Proceedings of the National Academy of Sciences of the United States of America, 98, 7641- .http://dx.doi.org/10.1073/pnas.121590798

Wendorf, F. (1968). The History of Nubia. Dallas, TX.

Winters, C. (2008). Aurignacian Culture: Evidence of Western Exit for Anatomically Modern Humans. South Asian Anthropologist, 8, 79-81.

Winters, C. (2010). Origin and Spread of the Haplogroup N. Bioresearch Bulletin, 3, 116-122.

Winters, C. (2011). The Gibraltar out of Africa Exit for Anatomically Modern Humans. WebmedCentral BIOLOGY, 2, Article ID: WMC002311. http://www.webmedcentral.com/article_view/2311

Notes:

1. C. Winters,A GENETIC CHRONOLOGY OF AFRICAN Y-CHROMOSOMES R-V88 AND R-M269 IN AFRICA AND EURASIA,http://www.cibtech.org/J-LIFE-SCIENCES/PUBLICATIONS/2017/VOL-7-NO-2/04-JLS-004-WINTERS-A-EURASIA.pdf

2. Lahovary N. (1963). Dravidian Origins and the West. Madras: Longmans.

3. MacWhite,E. (1947). Studios sobre las relaciones atlanticas de la peninsula hispanica en la edad del bronce. Dissertationes Matritenses, Vol.12.

4. Fregel R, et al (2017). Neolithization of North Africa involved the migration of people from both the Levant and Europe. bioRxiv 191569; doi: https://doi.org/10.1101/191569

5. Martín-Socas D, et al. (2004).Cueva de El Toro

(Antequera, Málaga-Spain). A Neolithic Stockbreeding Community in the Andalusian region between VI-III millenniums B.C.. Documenta Praehistorica XXX:126-143 . [accessed Sep 22, 2017]. Available from https://www.researchgate.net/publication/309558230_Cu eva_de_El_Toro_Antequera_Malaga-Spain_A_Neolithic_Stockbreeding_Community_in_the_A ndalusian_region_between_VI-III_millenniums_BC

6. Müller, J. & van Willigen, S. (2001). New radiocarbon evidence for European Bell Beakers and the consequences for the diffusion of the Bell Beaker phenomenon. In Bell beakers today: pottery, people, culture, symbols in prehistoric Europe. Proceedings of the International colloquium, Riva del Garda Trento, Italy, (ed. Nicolis, F., pp.59–80) .

7. Olalde I, Selina Brace, Morten E. Allentoft, Ian Armit, Kristian Kristiansen, et al. (2017). The Beaker Phenomenon and the Genomic Transformation of Northwest Europe. bioRxiv 135962; doi: https://doi.org/10.1101/135962

8. Cardoso, J. L. (2014).Absolute chronology of the Beaker phenomenon North of the Tagus estuary : demographic and social implications. TRABAJOS DE PREHISTORIA, 71, 56–75 .

9. Turek, J. 2012: Chapter 8 – "Origin of the Bell Beaker phenomenon. The Moroccan connection", In: Fokkens, H. & F. Nicolis (eds) 2012: Background to Beakers. Inquiries into regional cultural backgrounds of the Bell Beaker complex. Leiden: Sidestone Press. https://www.academia.edu/1988928/Turek_J._2012_Cha pter_8_-

10. Mathieson I, Songül Alpaslan Roodenberg, Cosimo Posth et al. (2017). The Genomic History of Southeastern Europe. http://biorxiv.org/content/early/2017/05/09/135616

Kushites lived in Anatolia and the Steppes

Ancient DNA (aDNA) indicates that R1 clades were carried by European hunter-gathers (CHG) and European farmers or early farmers (EF). Villabruna man lived 14kya in Italy and carried R1b1a. European hunter gatherers carried R1b1 in Spain and Samara. Many European farmers also carried varied R1 clades.

Although the lineages R1b1 and R1b1a were recognized as R-V88 clades (Cruciani et al, 2010), some researchers claim that Y-Chromosome R1 is of Eurasian origin, without any collateral evidence fromarchaeology to support this claim.
Controversy surrounds the presence of Y-Chromosome R1 in Africa, Cruciani et al., (2010) has suggested that the presence of R1 is the result of a back migration from Eurasia to Africa.

Haber et al., (2016) argues that a linkage-disequilibrium decay method indicated that there were two Eurasian migrations back to Africa. The researchers believe the first event occurred 4700-7200 years ago which resulted in a

Eurasian backflow to Africa of the Y-Chromosome lineage R1b-V88 by Neolithic LBK (Linearbandkeramik) or Linear Pottery population. The researchers claim that the LBK farmers probably migrated from the Near East to Northern Chad during the African Humid periods.

Haber et al, speculated that using PCA and MSMC analysis that the second Eurasian migration occurred 3kya. During the second migration the researchers claim, Eurasians deposited R-V88 among the Toubou, Laal and Sara who have 20%-34% R1b-V88 ancestry.

Haber et al., (2016) claim that Eurasian migrations back into Africa, explains the Neanderthal ancestry of ~0.5% in the Toubou and ~1.0% in the Amhara, while there was "no detectable genetic impact on other Chadian populations".

This theory lacks congruence because 1) there is no archaeological evidence of an Eurasian population in Africa (Winters, 2011,2014); 2) there is evidence of a Eurasian admixture in East and West Africa among populations that do not carry R-V88 or speak Afro-Asiatic languages (Pickrell,2014). Moreover, many African tribes like the Fulani , and Hausa carry R1b, but they are not from the Middle East (Winters,2010c).

Haber et al., (2016) claims that Eurasians passed on Neanderthal ancestry to the Toubou and Amhara during the second migration. The researchers statement relating to this matter is contradictory, because Haber et al., (2016) says that Eurasians took R-V88 to Africa, the idea that "Neanderthal ancestry to Africa "diluted" the Neanderthal ancestry in the Near East ", because the only way Neanderthal ancestry could have been "diluted" in the

Near East by a physical presence of Africans in the region during the Neolithic (2016). Occam Razor suggest that if there was a physical presence of Africans in the Near East to "dilute" Neanderthal ancestry there, Africans probably took R1b to Eurasians instead of a backflow from Eurasia (Winters, 2011,2014b).

In this chapter we are examining the Genomic structure of the populations that introduced the hunter-gather culture to Europe and its agro-pastoral traditions. We based our analysis on public genomic datasets and published articles. No new biological samples were collected for this study.

Analyzing the aDNA literature we assembled genome-wide data on the ancient populations of Africa, Mesopotamia and Crete. This data was used to obtain a profile of the haplogroups carried by the Kushites. The populations examined include the Early European Farmers (EF) , Anatolian Agro-Pastoralists (AAP) and the Caucasus Hunter Gathers (CHG).

Niger Congo-Dravidian Speakers

The linguistic, anthropological and linguistic data make it clear that the Dravidian people came to India from Africa during the Neolithic and not the Holocene period. Controversy surrounds the origin of the Dravidian languages. Winters (1989,2002) outlines the alleged relationship of Dravidian languages to Elamite, Sumerian and Japanese.

Although the relationship of Dravidian languages to these languages are disputed, there is abundant evidence that Dravidian languages are genetically related to the Niger-

Congo group (Aravanan 1979, 1980 ; Homburger 1948, 1957; sergent,1992; Upadhyaya and Upadhyaya, 1976, 1979; Winters 1985, 1988, 1989, 2002).

In the sub-continent of India, there were several main groups. The traditional view for the population origins in India suggest that the earliest inhabitants of India were the Negritos, and this was followed by the Proto-Australoid, the Mongoloid and the so-called Mediterranean type which represent the ancient Egyptians and Kushites (Winters 1985). The the Proto-Dravidians were probably one of the cattle herding groups that made up the C-Group culture of Nubia Kush (Aravanan, 1976; Winters, 2007, 2008).

B.B. Lal (1963) an Indian Egyptologist has shown conclusively that the Dravidians originated in the Saharan area 5000 years ago. He claims they came from Kush, in the Fertile African Crescent and were related to the C-Group people who founded the Kerma dynasty in the 3rd millennium B.C. (Lal 1963; Winters,1985,2002). The Dravidians used a common black-and-red pottery, which spread from Nubia, through modern Ethiopia, Arabia, Iran into India as a result of the Proto-Saharan dispersal (Winters, 2002, 2012).

B.B. Lal (1963) a leading Indian archaeologist in India has observed that the black and red ware (BRW) dating to the Kerma dynasty of Nubia, is related to the Dravidian megalithic pottery. Singh (1982) believes that this pottery radiated from Nubia to India. This pottery along with wavy-line pottery is associated with the Saharo-Sudanese pottery tradition of ancient Africa.

Aravaanan (1980) has written extensively on the African

and Dravidian relations. He has illustrated that the Africans and Dravidian share many physical similarities including the dolichocephalic indexes (Aravaanan 1980), platyrrhine nasal index (Aravaanan 1980), stature (31-32) and blood type (Aravaanan, 1980). Aravaanan (1980) also presented much evidence for analogous African and Dravidian cultural features including the chipping of incisor teeth and the use of the lost wax process to make bronze works of arts (Aravaanan 1980).

There are also similarities between the Dravidian and African religions. For example, both groups held a common interest in the cult of the Serpent and believed in a Supreme God, who lived in a place of peace and tranquility. There are also affinities between the names of many gods including Amun/Amma and Murugan. Murugan the Dravidian god of the mountains parallels a common god in East Africaworshipped by 25 ethnic groups called Murungu, the god who resides in the mountains.

There is physical evidence which suggest an African origin for the Dravidians. The Dravidians live in South India. The Dravidian ethnic group includes the Tamil, Kurukh,Malayalam, Kananda (Kanarese),Tulu, Telugu and etc. Some researchers due to the genetic relationship between the Dravidians and NigerCongo speaking groups they call the Indians the Sudroid.

Dravidian languages are predominately spoken in southern India and Sri Lanka. There are around 125 million Dravidian speakers. These languages are genetically related to African languages. The Dravidians are remnants of the ancient Black population who occupied most of ancient

Asia and Europe.

Linguistic Evidence

1.1 Many scholars have recognized the linguistic unity of Black African (BA) and Dravidian (Dr.) languages. These affinities are found not only in the modern African languages but also that of ancient Egypt. These scholars have made it clear that lexical, morphological and phonetic unity exist between African languages in West and North Africa as well as the Bantu group.

1.2 K.P. Arvaanan (1976) has noted that there are ten common elements shared by BA languages and the Dr. group. They are (1) simple set of five basic vowels with short-long consonants;(2) vowel harmony; (3) absence of initial clusters of consonants; (4) abundance of geminated consonants; (5) distinction of inclusive and exclusive pronouns in first person plural; (6) absence of degrees of comparison for adjectives and adverbs as distinct morphological categories; (7) consonant alternation on nominal increments noticed by different classes; (8)distinction of completed action among verbal paradigms as against specific tense distinction;(9) two separate sets of paradigms for declarative and negative forms of verbs; and (10) use of reduplication for emphasis.

1.3 There has been a long development in the recognition of the linguistic unity of African and Dravidian languages. The first scholar to document this fact was the French linguist L. Homburger (1950,1951,1957,1964). Prof. Homburger who is best known for her research into

African languages was convinced that the Dravidian languages explained the morphology of the Senegalese group particularly the Serere, Fulani group. She was also convinced that the kinship existed between Kannanda and the Bantu languages, and Telugu and the Mande group. Dr. L. Homburger is credited with the discovery forthe first time of phonetic, morphological and lexical parallels between Bantu and Dravidians.

ENGLISH	DRAVIDIAN	SENEGALESE	MANDING
MOTHER	AMMA	AMA, MEEN	MA
FATHER	APPAN, ABBA	AMPA,BAABA	BA
PREGNANCY	BASARU	BIR	BARA
SKIN	URI	NGURU, GURI	GURU
BLOOD	NETTARU	DERET	DYERI
KING	MANNAN	MAANSA, OMAD	MANSA
GRAND	BIIRA	BUUR	BA
SALIVA	TUPPAL	TUUDDE	TU
CULTIVATE	BEY	MBEY	BE
BOAT	KULAM	GAAL	KULU
FEATHER	SOOGE	SIIGE	SI, SIGI
MOUNTAIN	KUNDRU	TUUD	KURU
ROCK	KALLU	XEER	KULU
STREAM	KOLLI	KAL	KOLI

Figure 1: Common Niger-Congo-Dravidian Terms

1.6 By the 1970's numerous scholars had moved their investigation into links between Dr. and BA languages on into the Senegambia region. Such scholars as Cheikh T. N'Diaye (1972) a Senegalese linguist, and U.P. Upadhyaya (1973) of India, have proved conclusively Dr. Homburger's theory of unity between the Dravidian and the Senegalese languages.

1.7 C.T. N' Diaye, who studied Tamil in India, has identified nearly 500 cognates of Dravidian and the

Senegalese languages. Upadhyaya (1973) after field work in Senegal discovered around 509 Dravidian and Senegambian words that show full or slight correspondence.

1.8 As a result of the linguistic evidence the Congolese linguist Th. Obenga suggested that there was an Indo-African group of related languages. To prove this point we will discuss the numerous examples of phonetic, morphological and lexical parallels between the Dravidian group: Tamil (Ta.), Malayalam (Mal.), Kannanda/Kanarese (Ka.), Tulu (Tu.), Kui-Gondi, Telugu (Tel.) and Brahui; and Black African languages: Manding (Man.), Egyptian (E.), and Senegalese (Sn.).

Dravidian and Senegalese Cognates			
English	Senegalese	Dravidian	
body	W. yaram	uru	
head	D. fuko,xoox	kukk	
hair	W. kawar	kavaram 'shoot'	
eye	D. kil	kan, khan	
mouth	D. butum	baayi, vaay	
lip	W. tun,F. tondu	tuti	
heart	W. xol,S. xoor	karalu	
pup	W. kuti	kutti	
sheep	W. xar	'ram'	
cow	W. nag	naku	
hoe	W. konki		
bronze	W. xanjar	xancara	
blacksmith	W. kamara		
skin	dol	tool	
mother	W. yaay	aayi	
child	D. kunil	kunnu, kuuci	
ghee	o-new	ney	

Figure 2: Dravidian and Senegalese Cognates

6.1 Dravidian and Senegalese. Cheikh T. N'Diaye (1972) and U.P. Upadhyaya (1976) have firmly established the linguistic unity of the Dravidian and Senegalese languages. They present grammatical, morphological, phonetic and lexical parallels to prove their point.

6.2 In the Dravidian and Senegalese languages there is a tendency for the appearance of open syllables and the avoidance of non-identical consonant clusters. Accent is usually found on the initial syllable of a word in both these groups. Upadhyaya (1976) has recognized that there are many medial geminated consonants in Dravidian and Senegalese. Due to their preference for open syllables final consonants are rare in these languages.

6.3 There are numerous parallel participle and abstract noun suffixes in Dravidian and Senegalese.
For example, the past participle in Fulani (F) -o, and oowo the agent formative, corresponds to Dravidian -a, -aya, e.g., F. windudo 'written', windoowo 'writer'.

6.4 The Wolof (W) -aay and Dyolo ay , abstract noun formative corresponds to Dravidian ay, W. baax 'good', baaxaay 'goodness'; Dr. apala 'friend', bapalay 'friendship'; Dr. hiri 'big', hirime 'greatness', and nal 'good', nanmay 'goodness'.

6.5 There is also analogy in the Wolof abstract noun formative suffix -it, -itt, and Dravidian ita, ta, e.g., W. dog 'to cut', dogit 'sharpness'; Dr. hari 'to cut', hanita 'sharpness'.

6.6 The Dravidian and Senegalese languages use

reduplication of the bases to emphasize or modify the sense of the word, e.g., D. fan 'more', fanfan 'very much'; Dr. beega 'quick', beega 'very quick'.

6.7 Dravidian and Senegalese cognates (See: Figure 2). Above we provided linguistic examples from many different African Supersets (Families) including the Mande and Niger-Congo groups to prove the analogy between Dravidian and Black African languages (See: Figure 3) . The evidence is clear that the Dravidian and Black African languages should be classed in a family called Indo-African as suggested by Th. Obenga. This data further supports the archaeological evidence accumulated by Dr. B.B Lal (1963) which proved that the Dravidians originated in the Fertile African Crescent.

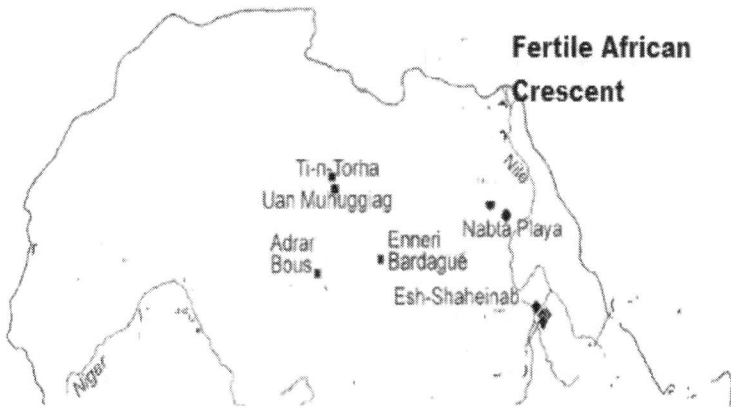

Using archaeological evidence we will reconstruct the set of migration processes that led to the raise of Neolithic cultures in Eurasia. The archaeological evidence indicates that Africans made several migrations into Eurasia. But, we have no archaeological evidence of a Eurasian culture carried back into Africa by any population (Winters, 2017b).

Jones et al., (2015), make it clear that "Given their geographic origin, it seems likely that CHG [Caucasus hunter-gatherers] and EF [European Farmers] are the descendants of early colonists from Africa who stopped south of the Caucasus, in an area stretching south to the Levant and possibly east towards Central and South Asia". We also see an influx of hunter-gathers and EF in Western Eurasia from Africa, via Iberia (Winters, 2017b).

The African origin of these Levantines is supported by the history of Kushites . Trenton W. Holliday (2000), tested the hypothesis that if modern Africans had dispersed into the Levant from Africa, "tropically adapted hominids" would be represented in the archaeological history of the Levant, especially in relation to the Qafzeh-Skhul hominids. Holliday (2000) found that the Qafzeh-Skhul hominids (20,000-10,000), were assigned to the Sub-Saharan population, along with the Natufians samples (4000 BP).

The Natufians and other Levantines carried haplohgroup E, which originated in Africa (Lazaridis ,2015) . Holliday (2000) also found African fauna in the area.

In recent years researchers have published work on the aDNA of Anatolia and Lower Egypt (Kilinç et al,2016; Martinez et al, 2007; Schuenemann et al, 2017), that allows us to present a fuller picture of Kushite genetic history. An examination of this history makes it clear that the Kushites, like other African population carried genes which have been misidentified as Eurasian.

Kushites

The Kushites belonged to the C-Group culture of Nubia. The Kushites spoke Niger-Congo and Dravidian languages (Winters,2012) (See: Supplementary File 1). The Niger-Congo (NC) Superfamily of languages is the largest family of languages spoken in Africa. Researchers have assumed that the NC speakers originated in West Africa in the Inland Niger Delta. The research indicates that the NC speakers originated in the Saharan Highlands 12kya and belonged to the Ounanian culture (Winters,2012).

The Ounanian culture is associated with sites in central Egypt, Algeria, Mali, Mauretania and Niger (Winters,2012). The Ounanian tradition is associated with the Niger-Congo phyla (Winters, 2012). This would explain the close relationship between the Niger-Congo and Nilo-Saharan languages.

The original homeland of the Niger-Congo speakers was probably situated in the Saharan Highlands during the Ounanian period. From here NC populations migrated into the Fezzan, Nile Valley and Sudan as their original homeland became more and more arid.

This was probably the ancient homeland of the Dravidians, Egyptians, Sumerians, Niger-Kordofanian Mande, Anatolian Kushites and Elamite speakers. We call this part of Africa the Fertile African Crescent (Jelinek,1985, Winters, 1995,1991). We call these people the Proto-Saharans (Winters,1985,1991). The generic term for this group in the ancient literatures was: Kushite.

The Kushites lived in Africa and Asia (Winters, 2017b, Winters, 2018). Origination of these diverse Kushite tribes in the ancient Sahara, explains the analogy between the Bafsudraalam languages . These languages include [B]lack [af]rican [su]merian [draa]vidian e[lam]:Bafsudraalam.

The Kushites lived in Africa and Asia (Winters, 2017b). Origination of these diverse Kushite tribes in the ancient Sahara, explains the analogy between the Bafsudraalam languages.

Common Bafsudraalam Terms

Language	Chief	city,village	black/burnt
Dravidian	cira.ca	uru	kam
Elamite	salu		
Sumerian	sar	ur	
Manding	sa	furu	kam 'Charcoal'
Nubian	sirgi	amr	uru-me
Semitic	sarr		ham
Ubaid	sar	ur	
Egyptian 'blackland'	sr	mer	kemit
Hausa	sarkl	birni	
Paleo-African	*Sar	*uru	*

The IAM [Early Neolithic Moroccans] people (Fergel et al., 2017), were nothing more than huntergatherer Kushites that had originally belonged to the Ounanian Culture (Winters, 2012, 2017b). The Ounanians, like their Kushite descendants were great archers and based their civilization on hunting using the bow, and limited cattle domestication

(Winters, 2012, 2017b).

The Ounanian culture was first described by Breuil in 1930 at Ounan to the south of Taodeni in northern Mali. Ounanian Points are suggested to be the hallmark of the some Epipaleolithic industries in the central Sahara, the Sahel and northern Sudan, and dated to the early Holocene. The original homeland of the Niger-Congo speakers was probably situated in the Saharan Highlands during the Ounanian period. From here NC populations migrated into the Fezzan, Nile Valley and Sudan as their original homeland became more and more arid.

Wavy-line pottery

In the Eastern Sahara many individual types of tanged and shouldered arrowheads occur on early Holocene prehistoric sites along with Green Saharan/Wavy-line pottery (Drake et al., ,2010; Vernet et al, 2007) . 'Saharo-Sudanese Neolithic' wavy-line, dotted wavy line and walking-comb pottery was used from Lake Turkana to Nabta Playa, in Tibestim , Mauritania, on into in the Hoggar, in Niger. This pottery evolved into the Beaker Bell ceramics.

The Ounanian culture was not isolated in Africa. It was spread into the Levant. As a result, we have in the archaeological literature the name Ounan-Harif point. This name was proposed for the tanged points at Nabta Playa and Bir Kiseiba .

Harifian is a specialized regional cultural development of the Epipalaeolithic of the Negev Desert. Harifian has close connections with the late Mesolithic cultures of Fayyum and the Eastern Deserts of Egypt, whose tool assemblage resembles that of the Harifian.

The tangled Ounanian points are also found at Foum Arguin . These points were used from Oued Draa, in southern Morocco, to the Banc d'Arguin and from the Atlantic shore to the lowlands of northwestern Sahara in Mauritania . We now have DNA from Ounanian sites in Morocco.

All the burials in Ifri n'Amr o'Moussa site IAM1-IAM7 , are devoid of any artifacts, except for an original funeral ritual, which consists of placing a millstone on the skull (5) . These burials were dated from 4,850 to 5,250 BCE, they carried U6, M1, T2, X and K (Fregel et al, 2017). This suggests that Africans were already carrying this mtDNA. The spread of the Ounanians to Harif in the Levant explains the presence of these Kushite clades in the Levant and Anatolia.

The first Kushites came to the Levant with Narmer.

Narmer was the first ruler to unify ancient Egypt. The reason we know Narmer was a Kushite is the fact that the bulla called this part of the Negev **ḫ3ts.t** ("Kush") or ḫ3s.tj ("Kushite"). and we find Narmer's name on jars and serekhs from excavations in Israel and Palestine , for example Tel Erani, Arad, 'En Besor, Halif Terrace/Nahal Tillah and more, we can assume that if he was recognized as ruler of the area he was also a Kushite (Levy et al,1997).

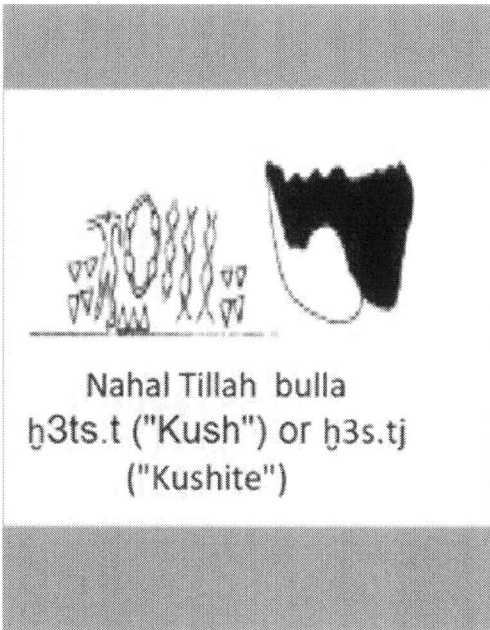

Nahal Tillah bulla
ḫ3ts.t ("Kush") or ḫ3s.tj ("Kushite")

The Kushites were called ḫ3št in Africa and the Levant. Kushites had early settled in the Levant since Narmer times. The Kushites were called ḫ3št , Ta-Seti and Tehenu by the Egyptians (Winters,2012,2017).

The Egyptian Pharaoh Sahure referred to the Tehenu leader as "Hati Tehenu" . The name Hati, correspond to the name Hatti for a Kushite tribe in Anatolia. The Hatti people often referred to themselves as Kashkas.

The Weni inscription from ancient Egypt acknowledges the fact that the Kushites lived not only in Nubia but, also in Lower Egypt, the Levant and Anatolia (Winters, 2017b). The Kushites living in Lower Egypt and across West Asia, were proud of their Kushite heritage and proudly declared their Kushite ancestry in the inscriptions of the Hyksos rulers of Lower Egypt, and the writings of the Hattians and Hurrians of Anatolia.

The Kushite tribes in Anatolia had many names including Kassite, Hurrian and Hattian. An important group in Anatolia in addition to the Hatti, were the Hurrians (Winters, 2018). The Hurrians entered Mesopotamia from the northeastern hilly area. They introduced horse-drawn war chariots to Mesopotamia.

The Kushites remained supreme around the world until 1400-1200 BC. During this period the Hua (Chinese) and Indo-European (I-E) speakers began to conquer the Kushites whose cities and economies were destroyed as a result of natural catastrophes which took place on the planet between 1400-1200 BC.

Later, after 500 AD, Turkish speaking people began to settle parts of Central Asia. Thie reason behind the presence of the K-s-h element in many place names in Asia e.g., Kashgar, HinduKush, and Kosh, is because these were regions settled by the Kushites. The HinduKush in Harappan times had lapis lazuli deposits.

Some of the Tehenu or Kushites settled Anatolia. The major Anatolian Kushite tribes were the Kaska and Hatti speakers who spoke the non-IE language called Khattili. The gods of the Hattic people were Kasku and Kusuh (< Kush).

The Hattic people are related to the Hatiu, one of the Egyptian Delta Tehenu tribes. Many archaeologist believe that the Tehenu people were related to the C-Group people. The Hattic language is closely related to African and Dravidian languages (Winters, 2014, 2017b,2018).

Hurrians

An important group in Anantolia in addition to the Hatti, were the Hurrians. The Hurrians entered Mesopotamia from the northeastern hilly area (Potts, 1995; Winters,2014). They introduced horse-drawn war chariots to Mesopotamia (Sagy, 1995).

Hurrians penetrate Mesopotamia and Syria-Palestine between 1700-1500 BC. The major Hurrian Kingdom was Mitanni, which was founded by Sudarna I (c.1550 B.C.), it was established at Washukanni on the Khabur River. The Hurrian capital was Urkesh, one of its earliest kings was called Tupkish.

Linguistic and historical evidence support the view that Dravidians influenced Mittanni and Lycia (Winters 2014). Alain Anselin is sure that Dravidian speaking peoples once inhabited the Aegean. For example Anselin (1982) and Winters (2014) has discussed many Dravidian place names found in the Aegean Sea area.

Two major groups in ancient Anatolia were the Hurrians and Lycians. Although the Hurrians are considered to be

Indo-European speakers, some Hurrians spoke a language related to the Dravidian languages (Winters,2014).

The Hurrians lived in Mittanni. Mittanni was situated on the great bend of the Upper Euphrates river. Hurrian was spoken in eastern Anatolia and North Syria.

Until recently, most of what we knew about Hurrian came from the Tel al-Armarna letters. These letters were written to the Egyptian pharaoh. The al-Armarna letters are important because they were written in a language different from diplomatic Babylonian. Other information on Hurrian comes from the Mitanni names in Akkadian and Sumerian (Wegner,1999; Winters, 2014).

The al- Armarna letters written in an unknown language to the Egyptians, were numbered 22 and 25. In 1909, Ferdinand Bork (1909), wrote a translation of the letters. Ilse Wegner (1999) used many examples from the Mitanni Letters in her discussion of Hurrian.

G.W. Brown (1930) proposed that the words in letters 22 and 25 were Dravidian especially Tamil. Brown (1930), has shown that the vowels and consonants of Hurrian and Dravidian are analogous. In support of this theory Brown (1930) noted the following similarities between Dravidian and Hurrian: 1) presence of a fullness of forms employed by both languages; 2) presence of active and passive verbal forms are not distinguished; 3) presence of verbal forms that are formed by particles; 4) presence of true relative pronouns is not found in these languages; 5) both languages employ negative verbal forms; 6) identical use of -m, as nominative; 7) similar pronouns; and 8) similar ending formations:

Dravidian Hurrian
  a   a
-kku -ikka
imbu impu

There are analogous Dravidian and Hurrian terms. They include kinship and cultural terms for example King, god, father and woman (Figure 4).

Comparison Hurrian and Dravidian Cultural Terms		
English	Hurrian	Dravidian
mountain	paba	parampu
lady, woman	aallay	ali
King	Sarr, zarr	Ca, cira
god	en	en
give	tan	tara
to rule	irn	ire
father	attai	attan
wife, woman	asti	atti

Figure 4: Hurrian and Dravidian Cultural/Kinship Terms

There are other Hurrian and Sanskrit terms that appear to show a relationship to the Tamil language as illustrated in Figure 5:

Comparison Hurrian, Sanskrit and Tamil			
English	Hurrian	Sanskrit	Tamil
One	aika	eka	okka 'together'
Three	tera	tri	
Five	panza	panca	añcu
Seven	satta	sapta	
Nine	na	nava	onpatu

Figure 5: Comparison Hurrian , Sanskrit and Tamil

Other Hurrian terms relate to Indo-Aryan as listed in Figure 6

Comparison Hurrian, Indo-Aryan and Tamil Terms			
Enlglish	Hurrian	I-A	Tamil
Brown	babru	babhru	pukar
Grey	parita	palita	paraitu 'old'
Reddish	pinkara	pingala	puuval
English	Mitanni	Vedic	Tamil
Warrior	marya	marya	makan, maravan

Figure 6: Hurrian Tamil and Indo-Aryan Terms

Although researchers believe that the Hurrians-Mitanni were dominated by Indo-Aryans (I-A) this is not supported by the evidence. Bjarte Kaldhol found that only 5 out of 500 Hurrian names were I-A sounding (Gupta, 2004).

The linguistic evidence discussed above is consistent with the view that the only Indian elements in Anatolian culture were of Dravidian, rather than Indo-Aryan origin. This evidence from Mittanni adds further confirmation to the findings of N. Lahovary in Dravidian Origins and the West, that prove the earlier presence of Dravidian speakers in

Anatolia. It also explains why we find Y-Chromosome R1a among the ancient Anatolians and European farmers.

The Hatti Using boats the Kushites moved down ancient waterways in Middle Africa and Arabia, many now dried up, to establish new towns in Asia and Europe after 3500 BC. The Kushites remained supreme around the world until 1400-1200 BC.

During this period the Hua (Chinese) and Indo-European (I-E) speakers began to conquer the Kushites whose cities and economies were destroyed as a result of natural catastrophes which took place on the planet between 1400-1200 BC. Later, after 500 AD, Turkish speaking people began to settle parts of Central Asia. **Aethiopians.**

This is the reason behind the presence of the K-s-h element in many place names in Asia e.g., Kashgar, HinduKush, and Kosh. The HinduKush in Harappan times had lapis lazuli deposits.

"A race divided, whom the sloping rays; the rising and the setting sun surveys…"

Most researchers assume that the ancient assertion of Kushites ruling the Middle East from Phoenicia to Syria is pure myth, however seals and other inscriptions of the Hyksos King Apophis suggest there may be some truth to the stories told by famous figures such as Homer and Strabo (Winters, 2017).

Around 800 BC, the Greek poet Homer mentions the, or Kushites, in the Iliad and the **Odyssey.** Homer said that the Kushites were "the most just of men, the favorites of the Gods".

To the Greeks and Romans there were two Kush empires, one in Africa and the other in Asia (Winters,2017). Homer alluded to the two Kushite empires when he wrote in the Odyssey i.23: "a race divided, whom the sloping rays; the rising and the setting sun surveys". In the Iliad. i.423, Homer wrote that Zeus went to Kush to banquet with the blameless Ethiopians.

In 64 BC, the Greek geographer and historian Strabo stated in Chapter 1 of Geography that there were two Kush empires - one in Asia and another in Africa. In addition to Kush in Nubia and Upper Egypt, some Greco-Roman authors considered their presence in southern Phoenicia up to Mount Amanus in Syria.

Kushites expanded into Inner Asia from two primary points of dispersal : Iran and Anatolia (Winters,2014) . In Anatolia the Kushites were called Hattians, Kassite, Elamites, Sumerians and Kaska.

Up until the 2nd millennium BC, the north and east of Anatolia was inhabited by non-I-E speakers. The Elamites lived in the Fars and the Bakhtiar valleys. This mountain area was named Elimaid in ancient times.

The Elamites called themselves: Khatan. The capital city of the Elamites Susa ,was called: Khuz by the Indo-European speakers, and Kussi by the Elamites. The Chinese called the Elamites Kashti. The Armenians called the eastern Parthia: Kushana.

Similar pottery was used in West Asia. The pottery from

Susa in Iran and Eridu in Mesopotamia of the fifth millennium BC are identical. Between 3700 and 3100 BC, Elam was under the influence of Uruk, as indicated by the shared art found at these sites during this period.

By the end of the 4th millennium BC , we see the beginnings of distinctive Elamite culture in the western Fars, at the Kur Valley. Here at Tel-i-Malyan we see the first Proto-Elamite tablets written in the Proto-Saharan script. Other Proto-Elamite writings soon appear at Susa.
The authors of the Proto-Elamite tablets were of Proto-Saharan origin. Malyan and Susa soon became the kingdoms of Anshan and Susa. These Proto-Elamites soon spread to Tepe Sialk and Tepe Yahya which was reoccupied after being abandoned earlier due to ecological decay.

The Proto-Saharans in Elam shared the same culture as their cousins in Egypt, Sumer, Elam and the Indus Valley. Vessels from the IVBI workshop at Tepe Yahya (c.2100-1700 BC), have a uniform shape and design. Vessels sharing this style are distributed from Soviet Uzbekistan, to the Indus Valley. In addition, as mentioned earlier we find common arrowheads at sites in the Indus Valley ,Iran, Egypt, Minoan Crete and early Heladic Greece.

Many of the Kushites in Anatolia belonged to the Tehenu Nation. Some of the Tehenu or Kushites settled Anatolia. Some of the major Anatolian Kushite tribes were the Kaska and Hatti speakers who spoke nonIE languages called Khattili. The gods of the Hattic people were Kasku and Kusuh (< Kush).

The Tehenu was composed of various ethnic groups. One of the Tehenu tribes was identified by the Egyptian

Pharoah as the Hatiu or Haltiu (El Mosallamy,1986). Some people claim that *ḫ3st* means 'foreign lands'. Semantically, for example "Wawat Rulers of foreign lands", is incorrect, because Wawat was the name of a nation, not a king. As a result, ḫ3st, was used to identify the nationality of the Wawat, Kau and other Kushite = ḫ3st. Thusly, the inscription of Weni reads: "His majesty made war on the Asiatic Sand-dwellers and his majesty made an army of many ten thousands; in the entire South, southward to Elephantine, and northward to Aphroditopolis [Busiris]; in the Northland on both sides entire in the [stronghold], and in the midst of the [strongholds], among the Irthet khas [Kusites], the Mazoi khas [Kushites], the Yam khas [Kushites], among the Wawat Khas [Kushites], among the Kau khas [Kushites], and in the land of Temeh " (Winters, 2018).

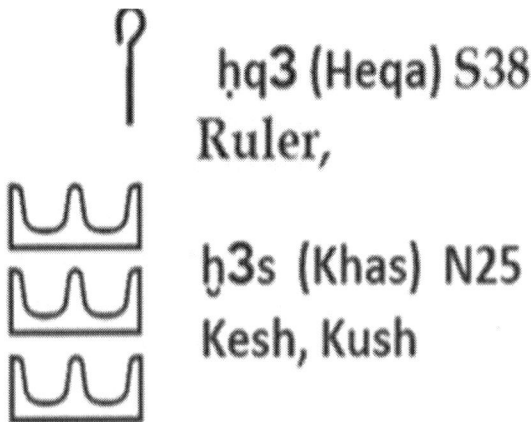

ḥq3 (Heqa) S38
Ruler,

ḫ3s (Khas) N25
Kesh, Kush

Figure 7: The Egyptian Signs for Kush

In the Weni inscription we can clearly see that the ḫ3st or Kushites were living in Upper and Lower Egypt. It is made clear that khas [Kushites] were also "in the land of Temeh".

The Egyptians made it clear that LOWER EGYPT was called : TAMEH , and UPPER EGYPT : TA SHEMA . Because the khas [Kushites], were living in Lower Egypt, when the Kings of khas [Kushites] took control of Egypt during the Hyksos period they were returning to the lands of their ancestors *Heqa khas* [Kushites] =Kings of the Kushites.

The Weni inscriptions includes Wawat, Yam and Temeh as ḫ3st or Kushite Nation. In the Weni inscription we can clearly see that Kushites were living in Upper and Lower Egypt. It is made clear that khas [Kushites] were also "in the land of Temeh".

The *khas* [Kushites] belonged to the C-Group people and lived in Upper and Lower Egypt between 3700-1300 BC and were called *Tmhw (Temehus)*. The *Temehus* were organized into two groups: the *Thnw (Tehenu)* in the North and the *Nhsj (Nehesy)* in the South.

During the Fifth Dynasty of Egypt (2563-2423), namely during the reign of Sahure there is mention of the Tehenu people. Sahure referred to the Tehenu leader "Hati Tehenu" (El Mosallamy,1986; Winters, 2017). These Hatiu, may correspond to the Hatti speaking people of Anatolia. The Hatti people often referred to themselves as Kashkas or Kaska.

Map of Ancient Anatolian Kushite Tribes

Another major Kushite group that ruled from Mesopotamia to northern India was the Kassites. The Kassites, occupied the central Zagros. The Kassites were also called *Kashshu*. This name agrees with Kaska, the name of the Hattians. P.N. Chopra, in 'The History of South India', noted that the Kassite language bears unmistakable affinity to the Dravidian group of languages.

Anatolia was divided into two lands "the land of Kanis" and the "land of Hatti". The Hatti were related to the Kaska people who lived in the Pontic mountains (See Map).

Hattians lived in Anatolia. They worshipped Kasku and Kusuh. They were especially prominent in the Pontic mountains. Their sister nation in the Halys Basin were the Kaska tribes. The Kaska and Hattians share the same names for gods, along with personal and place-names . The Kaska had a strong empire which was never defeated by the Hittites (Singer, 2007).

Singer (1981) has suggested that the Kaska, are remnants of the indigenous Hattian population which was forced northward by the Hittites. But at least as late as 1800 BC, Anatolia was basically settled by Hattians.

Anatolia was occupied by many Kushite groups, including the Kashkas and or Hatti (Singer, 2007). The Hatti , like the Dravidian speaking people were probably related to Niger-Congo speakers since they were a Kushite tribe.

The Hatti controlled the city state of *Kussara*. Kussara was situated in southern Anatolia.

The earliest known ruler of Kussara was Pitkhanas. It was his son Anitta (c. 1790-1750 BC) who expanded the Kussara Empire through much of Anatolia.

Many researchers get the *Hittites (Nesa)* mixed up with the original settlers of Anatolia called *Hatti* according to Steiner (1981). He wrote "[T]his discrepancy is either totally neglected and more or less skillfully veiled, or it is explained by the assumption that the Hittites when conquering the country of Hatti adjusted themselves to the Hattians adopting their personal names and worshipping their gods, out of reverence for a higher culture".

It is clear that the Anatolians spoke many languages including:Palaic, Hatti, Luwian and Hurrian, but the people mainly wrote in Neshumnili. The first nation to use this system as the language of the royal chancery were Hatti.

Comparison of African and Hattic Words			
•English	Hattic	Egyptian	Malinke (Mande language)
powerful	ur	wr'great,big'	fara
protect	$uh	swh	solo-
head	tup	tp	tu 'strike the head'
up,upper	tufa	tp	dya, tu 'raising ground'
to stretch put	pd	pe,	bamba
o prosper	falfat	--	find'ya
pour	duq	---	du 'to dispense'
child	pin	,pinu	den
Mother	na-a	--	na
lord	sa	--	sa
place	-ka	-ka	

Figure 8: Comparison African and Hattic Terms

Neshili, was probably spoken by the Hatti, not the IE Hittite. Yet, this language is classified as an IE langauge. Researchers maintain that the Hatti spoke 'Hattili' or Khattili "language of the Hatti", and the IE Hittites spoke

"Neshumnili"/ Neshili (Diakonoff and Kohl,1990). Researchers maintain that only 10% of the terms in Neshumnili is IE. This supports the view that Nesumnili may have been a lingua franca.

Itamar Singer (1981) makes it clear that the Hittites adopted the language of the Hatti . Steiner (1981) wrote that, " In the complex linguistic situation of Central Anatolia, in the 2nd Millennium B.C. with at least three, but probably more different languages being spoken within the same area there must have been the need for a language of communication or lingua franca (i.e., Neshumnili), whenever commercial transactions or political enterprises were undertaken on a larger scale" (Singer, 1981).

The Hattic people, were members of the Hatiu tribe, one of the Delta Tehenu tribes. The Tehenu people were related to the C-Group people. The Hattic language is closely related to African and Dravidian languages (See: Figure 8).

The languages have similar syntax Hattic le fil 'his house'; Mande a falu 'his father's house'. This suggest that the first Anatolians were Kushites, a view supported by the Hattic name for themselves: *Kashka , ḫ3st* or Kushite Nation.

The Hatti language which provided the Hittites with many of the terms Indo-Aryan nationalists use to claim and Aryan origin for the Indus civilization is closely related to African languages including Egyptians (Figure 9).

Comparison Hattic and African Languages			
Language	Big, mighty, powerful	protect, help	upper
Hattic	ur	$uh	tufa
Egyptian	wr	swh	tp
Malinke	fara	solo	dya, tu 'raising'
	Head	stretch (out) prosper	to pour
Hattic	tu	put falfalat	duq
Egyptian	tup	pd	
Malinke	tu 'strike head'	pe, bemba fin'ya	du

Figure 9: Shared African and Hattic Terms

The Malinke-Bambara and Hatti language share other cognates and grammatical features. For example,in both languages the pronoun can be prefixed to nouns, e.g., Hatti le 'his', le fil 'his house'; MalinkeBambara a 'his', a falu 'his father's house'. Other Hatti and Malinke- Bambara cognates include: Hattic b'la ka -ka Kaati, Malinke n'ye teke -ka ka, kuntigi 'headman'

Good hypothesis generation suggest that given the fact that the Malinke-Bambara and Hatti languages share cognate terms, Sumerian terms may also relate to Hatti terms since they were also Kushites. Below we compare a few Hatti, Sumerian and Malinke-Bambara terms (Figure 10).

Hatti, Sumerian and Malinke Bambara terms				
Language	Mother	father	lord,ruler	build, to set up
Hattic	na-a		ša	tex
Malinke	na	baba	sa	te
Sumerian	na 'she'	aba		tu 'to create'
	To pour	child, son	up, to raise	strength, powerful land
Hatti	dug	pin,pinu	tufa	ur -ka
Malinke	du den	dya,	tu	fara -ka
Sumerian	dub	peš	dul	usu ki

Figure 10: Hatti, Sumerian and Malinke Bambara terms

We now have ancient DNA (aDNA) from Africa. This aDNA allows us to determine the DNA carried by Africans during the Iberomaurusian period and later . This aDNA is from Iberomaurusian skeletons exhumed from the archaeological site of Afalou (15,000–11,000 YBP) in Algeria, and the archaeological site of Taforalt (23,000–10,800 YBP) in Morocco (Kefi et al, 2016).

The researchers found five different mtDNA haplogroups: H, U, J, J1c3f and T2 (Kefi et al, 2016). Van de Loosdrecht et al, (2018), found that Taforalt population carried haplogroups M1b and U6. This makes it clear that as early as 10.8kybp-23kybp Africans were carrying mtDNA haplogroups: H, U, J, J1c3f. M1b, T2 and U6. The Y-chromosome among the Taforalt population was haplogroup E1b1ba1 (M-78) (Loosdrecht et al,2018) .

The Kushites expanded from Nubia, into Crete and West Asia. As a result, we find that the Cretans and Anatolians shared the same DNA. The Cretans were called Keftiu .

The Keftiu , are descendants of the Garamantes, a Mande speaking tribe.

The mtDNA haplogroups L1, L2, L3, M1, N, H, U5 and U6 are associated the Kushite speakers. Phylogenetically all the Eurasian mtDNA branches descend from haplogroup L3.

The highest concentration of U5, is found among the Berbers/Taureg in North Africa. It is also carried by Mande and Fulani speakers. The Djola Mande speakers, also carry mtDNA M1, H and N (Rosa et al,2009). The U5 haplogroup is characterized by hyplotypes 16189, 16192 , 16270 and 16320.
The Pan-African haplotypes are 16189,16192,16223, 16278,16294, 16309, and 16390. This sequence is found in the L2a1 which is highly frequent among the Mande speaking group and the Wolof.

The Minoan mtDNA was H (43.2%), T (18.9%), K (16.2%) and I (8.1%). Haplogroups U, W, J2, X and J, were each identified in a single individual.

Because the Cretans were Mande speakers we expect contemporary Mande to carry these genes. The Mande speakers, in the genetics literature are represented under varying names including Djola and Mandekan. The Djola and Mandekan carry 2% Eurasian admixture. Some of the Mande speakers live in Mali, and carry the N and H haplogroups

The genomic data for the ancient Cretans and Anatolians is important because in addition to Kushites living in Crete and Anatolia, Kushites also lived in Lower Egypt at Abusir

, Egypt (Schuenemann et al,2017).

In Schuenemann et al., (2017) there were 100 mummies in the study. A total of 27 mummies were dated between 992-749 BC. The dates for these mummies precede the entrance of Romans, Greeks, Turks and Arabs in Egypt.

As a result, the Abusir mummies dating between 992-749 BC, reflect the Kushites who lived in Lower (18.5%), X (0.07%), H (0.07%) and M1a (0.07). The presence of these haplogroups at Abusir, shows that the U,T and J clades had a high frequency among ancient Egyptians and Kushites. This is not surprising because the Iberomaurusians carried identical genes.

The aDNA of the Abusir mummies , as African Kushite DNA , is supported by recent aDNA from Ifrin'Amr o'Moussa . The Ifri n'Amr o'Moussa site is an Early Neolithic Moroccan site. The DNA, haplotypes from this site is dated between 4,850 to 5,250 BCE. The aDNA recovered was U6, M1, T2, X and K (Fregel et al, 2017).

The correspondence between the Iberomaurusian, Early Neolithic Moroccan DNA, and Abusir DNA , is due to the spread of the Ounanians [Kushites] to Harif in the Levant. The Ounan-Harif points, Weni text and identification of Hattic and Kaska tribes in Anatolia, explains the presence of these Kushite clades in the Levant and Anatolia (Winters,2018).

There is mtDNA data uniting Africans and Dravidians. Some researchers attempt to portray the Dravidians as Caucasoid people and try to link these people to western Eurasian populations. Other researchers in India attempt to postulate an Indian origin for Dravidians because they mainly belong to the M haplogroup (HG) (Kivisild et al, 1999; Thangaray et al, 2006).

Clearly, the dates for L3(M,N) in western Eurasian are incongruent to TMRCA of the populations carrying the L3(M,N) lineages into eastern Eurasia which probably date to 60-65kya. This incongruence in relation to the dates for this haplogroup in eastern Eurasia, and its complete absence in much of western Eurasia today suggest that the population carrying these genes into Eurasia may not have entered Eurasia during the recognized Africa exit event.

In addition to ancient mtDNA M1 in Africa, we also find haplogroups M*, M23, M3 positions 482 and 16126; M30 positions 195A and 15431; and M33 position 2361 (Winters,2006). It is interesting to note that the presence of these genes, which are normally found in India are also found in Africa, is interesting given the presence of M1 in India and the existence of these genes among populations stretching from Africa into Yemen on into India along a path associated with the spread of the Tihama culture (Winters, 2008) .

In addition to haplogroups M1, M* and N in Sub-Saharan Africa we also find among the Senegambians hapotype AF24 (DQ112852) , which is delineated by a DdeI site at 10394 and AluI site of np 10397. The AF-24 haplotype is a branch of the African subhaplogroup L3 . This is the same delineation of haplogroup M*. It is clear from the molecular evidence that the M1, M and N haplogroups are found not only in Northeast Africa, but across Africa from East to West (Winters, 2007).

Neanderthals dominated the Levant when the imagined

back migration of M1 occurred 50kya ,we must reject the contention of Gonzalez et al., (2007), Olivieri et al., (2007) and Sores et al., (2012) that M1 originated in Asia because 1) the possible Senegalese origin of the M1c subclade; 2) the absence of the AF-24 haplotype of haplogroup LOd in Asia; and 3) the African origin of the Dravidian speakers of India (Winters, 2007,2008) who carry the most diverse M haplogroups.

Moreover, the existence of the L3a(M) motif in the Senegambia characterized by the DdeI site np 10394 and AluI site np 10397 in haplotype AF24 (DQ112852) make a 'back migration of M1 to Africa highly unlikely, because of the ancientness of this haplotype and presence of the haplogroup among the Iberomaurusian (Loosdrechtet al, 2018),. The first anatomically modern human (amh) to reach Senegal belonged to the Sangoan culture which spread from East Africa to West Africa probably between 100-80kya.

Most researchers make it appear that the M1 haplogroup is only found in Ethiopia. These researchers maintain that the M1 HG is restricted to the Afro-Asiatic linguistic phylum. This is false M HGs are found in other parts of Africa where people speak non-Afro-Asiatic languages.

The M lineages are not found only in East Africa. Rosa et al., (2009) found a low frequency of the M1 HG among West Africans who speak the Niger Congo languages, such as the Balanta-Djola. Gonzalez et al., (2006) found N, M and M1 HGs among Niger-Congo speakers living in Cameroon, Senegambia and Guinea Bissau.

Thangaraj et al., (2009) recognize a Paleolithic origin for the

M haplogroups in India. The majority of Dravidian speaking people belong to the M haplogroup. Most geneticists agree that the M macrohaplogroups are derived from L3. Kivisild et al., (1999) made it clear that all Indian mtDNA lineages "coalesce finally to the African L3a".
Metspalu et al., (2004) argues that the earliest offshoots for L3, were HGs M and N developed in Arabia. Metspalu et al., (2004) believes the MRCA for the M HG entered Asia 60-65 kya .

Metspalu et al., (2004) maintains that "all the basal trunks of M, N and R have diversified in situ" (p.24). He makes it clear that in his opinion the M HGs are different from the subhaplogroup M of East Asia .

The most frequent HG in India is M2. Sixty percent of of the Indian mtDNA lineages are M HGs . Kivisild et al., maintains that there are five M HGs in India: M1, M2,M3, M4, and M5. Thanaraj et al., (2009) has revised the classification of HGs M3, M18 and M31 and defined the novel HG M41.

The diversity of M HGs in India has led many researchers to suggest that the M clades have an in-situ origin .These researchers speculate that although L3 originated in Africa, the M1 HG in Ethiopia and Egypt ,may be the result of a back migration to Africa from India . This theory can no longer be supported given the presence of M1 among the Iberomaurusians (Loosdrechtet al, (2018).

These researchers base this theory for a back migration to Africa from India, on 1) HG M1 is not found in India; and 2) the M HGs are only found in East Africa . Both of these theories have little support when we look at the mtDNA data for Africa and India. Barnabas et al., (2005) noted that N,M and F lineages found in India could have originated in Africa . He speculated that these people migrated to India from Africa during the Upper Paleolithic.

It is also not true that HG M1 is absent in India. Kivisild et al., (1999) found five M HGs in India: M1, M2, M3, M4 and M5. It is interesting to note that the M4 HG has the same 16311 coding region as the African M1 HG.

Kivisild et al., (1999) provides the first detailed discussion of the M subclusters in India and suggested an autochthonous development of these lineages in India. The researchers suggest that there were multiple M lineages when this haplogroup migrated to Asia (Guilaine,1976). These researchers claimed that the expansion date for the five M subclusters expanded into India between 17,000-32,000 before the present (bp).

Kivisild et al., (1999) noted that 26 of the subjects in his study belonged to the M1 haplogroup. It is clear from this study that sub-cluster M1 was found mainly in the Indian states of Kerala and Karnataka (Guilaine,1976). An interesting finding in the study was that most of the Indians with the M1 HG were members of upper caste. Africans and Dravidians share haplogroups M1, M3, M30 and M33 (Winters, 2006).

The expansion of Kushites in Asia explains why Africans carry so-called Eurasian genes (Winters,2017b). This proves that the back migration theory for L3(M,N) and Y-Chromosomes R, and G, originated in Africa thousands of years before anatomically modern humans exited Africa.

The phylogeography of Y-Chromosome haplotypes shared among the Kushite Niger-Congo speakers include A,B, Elb1a, E1b1b, E2, E3a and R1 . The predominate Y-Chromosome among the Niger-Congo is M2, M35, and M33.

Haplogroup E has three branches carried by Kushite populations E1, E2 and E3. The E1 and E2 clines are found exclusively in Africa. Haplogroup E3 is also found in Eurasia. Haplogroup E3 subclades are E3b, E-M78, E-M81 and E-M34.

The majority of Kushites belong to E1b1a, Elb1b, E2 and R1. Around 90% belong to Y-Chromosome group E (215, M35*).

Y-Chromosome haplogroup A is represented among Niger-Congo speakers. In West Africa, under 5% of the NC speakers belong to group A. Most Niger-Congo speakers who belong to group A are found in East Africa and belong to A3b2-M13: Kenya (13.8) and Tanzanian (7.0%).

The Bantu expansion is usually associated with the spread of Y-Chromosome E3a-M2. In Kenya the frequency for E3a-M2 is 52%; and 42% in Tanzania. In Burkina Faso high frequencies of E-M2* and EM191* are also represented. It is interesting to note that among the Mande speaking Bisa and Mandekan there are high frequencies of E-M2*. This is in sharp contrast to the Marka and South Samo who have high frequencies of E-M33.

Y-chromosome R1 is found throughout Africa. The pristine form of R1*-M173 is only found in Africa

(Cruciani et al, 2010, 2010b; Coia et al, 2005; Winters, 2010,2011) . The age of Y- Chromosome R is 27ky (Kivisild, 2017). There is a great diversity of the macrohaplogroup R in Africa as illustrated in Table 1.

The name for African R Y-Chromosome haplogroups in Africa are constantly being changed. In Figure 11, we see that in 2010, a predominant R Y-Chromosome clade in Africa is haplogroup R1b (Carvalho et al,2011; Cruciani et al., 2010, 2010b; Winters, 2016, 2017a) and R1b1 (Berniell-Lee at al,2009) . Cruciani et al., (2010) discovered new R1b mutations including V7, V8, V45, V69, and V88.

Geography appears to play an insignificant role in the distribution of haplogroup R in Africa. Cruciani et al., (2010) has renamed the R*-M173 (R P-25) in most of Africa V88. The TMRCA of V88 was 18 kya (Kivisild, 2017).

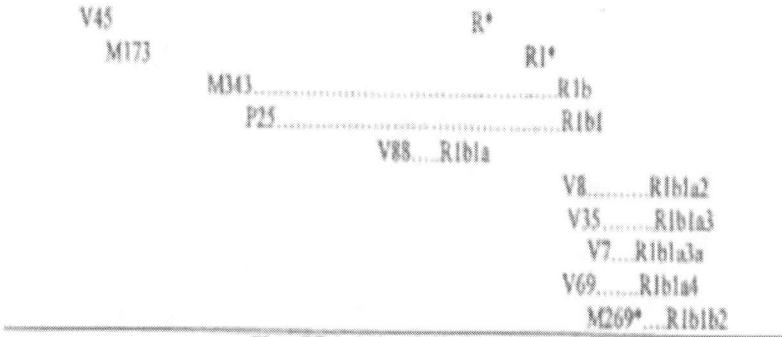

Figure11: African Y-Chromosomes 2010

Y-chromosome V88 (R1b1a) has its highest frequency among Chadic speakers, while the carriers of V88 among Niger-Congo speakers (predominately Bantu people) range between 2-66% . Haplogroup V88 includes the mutations M18, V35 and V7. Cruciani et al., (2010) revealed that R-V88 is also carried by Eurasians including the distinctive mutations M18, V35 and V7. Haplogroup R1b1-P25 was originally thought to be found only in Western Eurasia. Haplogroup R1b1* is found in Africa at various frequencies. Today R1b1 is called R-L278.

The first offshoot of R1b-M343 was V88. The Y-Chromosome V88 is a signature African haplogroup.

Toomas Kivisild (2017) noted: "Interestingly, the earliest offshoot of extant haplogroup R1b-M343 variation, the V88 sub-clade, which is currently most common in Fulani speaking populations in Africa (Cruciani et al, 2010; Winters, 2010b, 2011), has distant relatives in Early Neolithic samples from across wide geographic area from Iberia, Germany to Samara ." The relative of V88 in ancient Europe was R1b1.

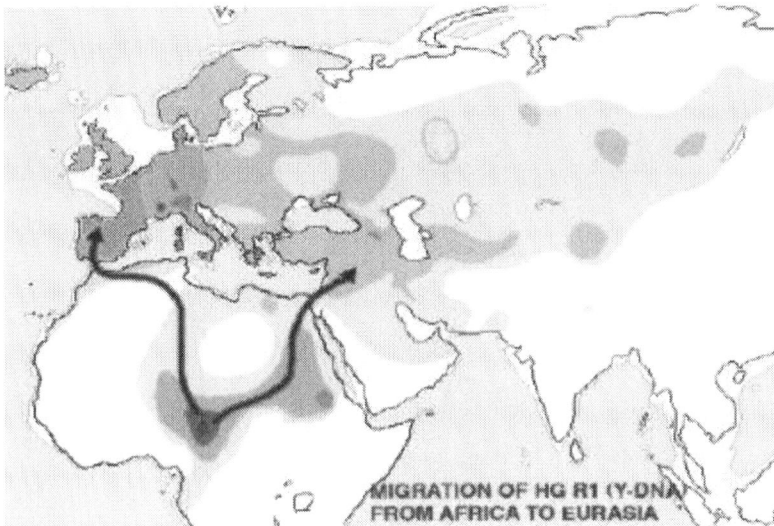

MIGRATION OF HG R1 (Y-DNA) FROM AFRICA TO EURASIA

In 2010, R-V88 was originally named R1b1a . Today R-V88 is named R1b1a2, and R1b1a is renamed RL754.

The ancient Europeans and Africans share R-L278 and R-L754. The earliest carrier of R-L278 in Europe was the hunter-gatherer Villabruna man in Italy. Villabruna man lived 14kya. We also had huntergatherers carrying RL278 (R1b1) in Spain and Samara. This would place Africans carrying R-L278 in Europe long before the origination of the Bell Beaker and Yamnaya cultures.

The Kushite haplogroups in Crete and West Asia varied. The Y-Chromosome among the Cretans and Anatolians were J,G, R1a1, R1b, T, K and H.

Martinez et al., (2007), observed that in the case of the R1 haplogroup, while frequencies of 19.2% and 21.7% are found in the Heraklion Prefecture and Lasithi Prefecture populations, respectively, more than half (56.1%) of the Lasithi Plateau individuals are R1-M306-derived.

In the case of Cretan E3b3-M123 (M34) chromosomes, they most likely signal East African or MiddleEastern gene flow rather than European, due to the scarcity of this lineage in the latter area. Similarly, the presence of E3b-M35* individuals in the Heraklion Prefecture population could probably be attributed to an East-African or North-African contribution.

The finding that other Minoans carried haplotype T and K also indicates that the Minoans were Blacks,not whites. There are a number of shared African and Indian Y-chromosome haplotypes. These haplotypes include Y-HG T-M70 and H1. Haplogroup T-M70 is found among several Dravidian speaking tribal groups in South India, including the Yerukul (or Kurru) , Gonds and Kols. Y-haplogroup T-M70 is found in the eastern and southern regions of India (Trivedi et al, 2008). It has a relatively high frequency in Uttar Pradesh and Madhya Pradesh (Sharma et al, 2009).

Sharma et al., (2009) in a study of 674 Dalits found that 89.39 % belonged to Y-HG K*, in relation to Dravidian speakers it was revealed that Y-HG TM70 was 11.1%. Trevedi et al., (2008) report that Y-HG T-M70 is predominately found among Upper Caste Dravidians at a frequency of 31.9. The highest frequency of T-M70 in the World is found among the Fulani (18%) of West Africa. Martinez et al., (2007) also found T-M70 and hg K in Crete .

Ramana et al., (2001) claims that the discovery of H1 and H2 haplotypes among the Siddis is a "signature" of their African ancestry. As a result, the Y-HG H1 subclade frequency among Dravidian speakers can also be considered as an indicator of an African-Cretan-Dravidian connection.

The H1 haplotype is found among many Dravidians (Winters,2010d). Sengupta et al., (2006) noted that the subclades H1 and H2 were found among 26% of the Dravidian speakers in their study, especially in Tamil Nadu. Trivedi et al., (2008) found the Y-hg H1 frequency of 22.2 among Dravidian speakers in their study. Sharma et al., (2008) reports a frequency rate of 25.2%.

We looked at previously published Y-Chromosome haplotype gene variants in Indian populations. The H1 haplotype is found among many Dravidians. Sengupta et al., (2006) noted that the subclades H1 and H2 was found among 26% of the Dravidian speakers in their study, especially in Tamil Nadu . Ramana et al., (2001) claims that the discovery of H1 and H2 haplotypes among the Siddis is a "signature" of their African ancestry (Winters, 2010e). The frequency of the H1 subclade among Dravidian speakers is also an indicator of an African-Dravidian connection (Winters, 2007, Winters, 2008, Winters, 2010d, Winters,2010e).

Watkins et al., (2008) noted that the common Indian Y-haplogroups were predominantly R1a1 (27%) and R2 (11%)); the major lineage in the Tamil castes, include H (21%, predominantly H1), L (13%, predominantly L1), J (11%, predominantly J2), and F* (10%).
In addition to haplotypes H1, in South India we also find the Sickle Cell gene (Winters, 2010e) and African 9-bp deletion (Winters, 2010d). Watkins et al., (2008) found the 9bp motif among four Indian tribal populations: Irula, Yanadi, Siddi and Maria Gond (Winters, 2010e).

Figure 12:Ancient DNA Frequencies European Cultures

Culture	mtDNA		Y- Chromosome	
	H	U5	R1a	R1b
Yamna	21%	13%	0%	91.50%
Corded Ware	21%	13%	71%	11.50%
German Bell Beaker	25%	14%	0%	100%

The phylogenetic structure of the Dawoodi Bohra Muslims of Tamil Nadu, India includes African mtDNA and Y-chromosome genes. The Dawodi Bohra carry the mtDNA M1 and Loa2a. The African Ychromosomes found among the Dawoodi Bohra was 20% haplotype H and 2% E1b1b1a. This evidence makes it clear that many of the so-called Eurasian clades were in Lower Egypt, before the Greco-Romans, Turks and etc., ruled Egypt. And as a result, Y-Chromosomes R.J, and G; and mtDNA U,M,T, J and N clades were Kushite lineages. This evidence makes it clear that the Kushites took R1a and R1b to Europe. Using samples from the aDNA literature allows us to determine the frequencies of mtDNA and Y-Chromosome clades carried by the ancient Europeans (Haak et al., 2015; Haber et al,2016; Kivisild, 2017; Mathieson et 2015, 2017 ;Olalde et al,2017) (See Figure 12).

Cattle Domestication

Agro-Pastoral Kushites cultivated crops and herded cattle. Elements of the Agro-Pastoral members of the Bell Beaker and Corded Ware complexes appear first in the African Sahara. Here we see rock engravings of cattle herders and hunters using similar bow and arrows.The Yamnaya archers' wrist-guard and bows may have had their origin in the Sahara where we see similar wrist-guards (Quellec, 2011).

The Niger-Congo speakers or Kushites formerly lived in the highland regions of the Fezzan and Hoggar until after 4000 BC. Originally hunter-gatherers the Kushites developed an agro-pastoral economy which included the cultivation of millet, and domestication of cattle (and sheep).

As early as 15,000 years ago cattle were domesticated in Kenya. In the Sahara-Nile complex, people domesticated many animals including the pack ass, and a small screw horned goat which was common from Algeria to Nubia.
The zebu or humped cattle are found in many parts of Africa. We find rock art depicting humped cattle dating back to 7000 BP. The oldest faunal remains of the Bos Indicus come from Kenya, and date to the first millennium B.C.

The recent evidence that Bos Indicus , humped cattle, may have originated in East Africa suggest that this type of cattle may have first been situated in Africa, and then taken to Asia by the Proto-Saharans. This view is supported by the fact that the advent of the Bos Indicus, cattle in Egypt corresponds to the migration of the C-Group people into the Nile Valley.

The C-Group people came from the Fertile African Crescent. Pastoralism was the first form of food production developed by post Paleolithic groups in the Sahara.

In the western Saharan sites such as Erg In-Sakane region, and the Taoudenni basin of northern Mali, attest to cattle husbandry between 6000 and 5000 B.P. Cattle pastoral people began to settle Dar Tichitt and Karkarchinkat between 5000 and 3500 B.P.

At Nabta Playa the people herded cattle and cultivated crops. The Kushites cultivated pennisetum millet at Nabta Playa (c. 7950 BC) and probably herded cattle (Miller, et al, 2010; Mitchell,2013).

During the Ounanian period, due to abundant fertility in the Sahara , many people herded cattle. Nabta Playa was located on the shoreline of a lake 11,000 years ago.

A center of cattle worship was the Kiseiba -Nabta region in Middle Africa. At Nabta archaeologists have found the oldest megalithic site dating to 6000-6500 BC, which served as both a temple and calendar.

This site was found by J. McKim Malville of the University of Colorado at Boulder and Fred Wendorf of Southern Methodist University.

Millet collection/cultivation

The Kushites introduced the agro-pastoral tradition to Europe and Anatolia. The Kushites took with them their cattle and Pennisetum millet.

Many of the Niger-Congo and Dravidian speaking Kushites carried R1b and R1a clades. The millet cultivated by the European farmers had to come from Africa, because millet was not cultivated in Central Asia until after it had been established in Anatolia and Europe.

The major grain exploited by Kushites were rice ,the yam and Pennisetum millet . The principal domesticate in the southern Sahara was bulrush millet.

Archaeologists have found charred millet remains at Nabta Playa (c.7950 BC). Millet impressions have been found on Mande ceramics from both Karkarchinkat in the Tilemsi Valley of Mali, and Dar Tichitt in Mauritania dating between 4000 and 3000 BP (McIntosh and McIntosh,1983; Winters,1986).

The earliest date for millet cultivation is not Dadiwan China (c.5000BC), it is Nabta Playa (c.7950 BC).It would appear that millet came to Europe with the CHG, and EF populations. Researchers have been troubled by the fact that although millet collecting began at Sokoltsy, Ukraine (c.7440 BC) (Hunt et al, 2008) with the CHG , there is no evidence of millet cultivation in Central Asia until the end of 3rd millennium BC (Hunt el al, 2008).

The presence of millet at Sokoltsy, indicates that millet cultivation in Europe, did not come from China or Central Asia. The early date of millet cultivation in Ukraine indicates that the cultivators of millet in Europe came from Africa.

The cultivation of millet at Nabta Playa, four hundred years before its cultivation at Sokoltsy indicates that Africans carrying R1 had left Nabta Playa and began cultivating millet in Eastern Europe. This would explain the presence of R1b1 in Italy and Samara among the CHG (Haak et al, 2015; Kivisild, 2017).

Cultivation of Pennisetum millet moved from East to West across Europe (Miller et al,2010). The earliest millet is found impressed on pottery at the Sokoltsy 2 Ukraine site , dating to around 6300 BC (Hunt et al,2008). This indicates that European hunter-gatherers were collecting millet after the Kushites were collecting millets in Africa at Nabta.

There is an absence of millet cultivation in Central Asia prior to the discovery millet in the mid-2nd millennium BC, at Begash, Kazakhstan (2300-2100BC). Millet has been found at Tahirbaj Tepe , this indicates the first evidence of the cultivation of millet in Anatolia and Europe. Motuzaite-Matuzeviciute et al., (2013) has argued that cultivation of broomcorn millet did not become widespread in the Caucasus, Iran and Syria until during the Bronze Age. Millets were cultivated at Shortugai, Afghanistan, and Ojakly, Gonur in the Murghab delta. The Anatolian sites include Gordion , Haftavan and Kilise (Miller et al,2010).

Kushites also took millet to India (Winters,2008). There has been considerable debate concerning the transport of African millets to India (Winters,1980,2008). Weber (1998) believes that African millets may have come to India by way of Arabia . Wigboldus(1996) on the other hand argues that African millets may have arrived from Africa via the Indian Ocean in Harappan times (Winters, 1980,2008).

Both of these theories involve the transport of African millets from a country bordering on the Indian Ocean. Yet, Weber (1998) and Wigboldus (1996)were surprised to discover that African millets and bicolor sorghum, did not reach many East African countries until millennia after they had been exploited as a major subsistence crop at Harappan and Gujarat sites (Winters,1980).

This failure to correlate the archaeological evidence of African millets in countries bordering on the Indian Ocean, and the antiquity of African millets in Africa, Anatolia and India suggest that African millets such as Pennisetum and Sorghum must have come to India from another part of Africa. That place from whence they came had to have been Middle Africa, given the antiquity of millet collection at Nabta Playa.

The Neolithic British farmers were genetically similar to Neolithic Iberians farmers dating between 3900–1200 BCE (Lahovary,1963; MacWhite,1947; Olade et al., , 2017) . The British farmers were replaced by farmers of the Beaker culture (Lahovary,1963). Eighty-four percent of the Beaker Bell Steppe migrants carried R1b (Olade et al, 2017).

One of the principal groups to use millet in Africa are the Northern Mande speaking people (Winters, 1986). The Northern Mande speakers are divided into the Soninke and Malinke-Bambara groups. The founders of the Dhar Tichitt site where millet was cultivated in the 2nd millenium B.C., were northern Mande speakers.

To test this theory we compared Dravidian and Black African agricultural terms, especially Northern Mande (See: Figure 4). The linguistic evidence suggest that the Proto-Dravidians belonged to an ancient sedentary culture which existed in Saharan Africa. We will call the ancestor of this group Paleo-DravidoAfricans.

The Dravidian terms for millet are listed in the Dravidian Etymological Dictionary at 2359, 4300 and 2671. A cursory review of the linguistic examples provided below from the Dravidian, Mande and Wolof languages show a close relationship between these languages. These terms are outlined below in Figure 13.

Dravidian and African Terms for Millet				
Kol	sonna	---	---	----
Wolof (AF.)	suna	---	----	---
Malinke (AF)	suna	bara, baga	de-n, doro	koro
Tamil	connal	varaga	tinai	kural
Malayalam	colam	varaku	tina	---
Kannanda	---	baraga, baragu	tene	korale,korle
	*sona	*baraga	*tenä	*kora

Figure 13: Dravidian and African Terms for Millet

It is clear that the Dravidian and African terms for millet are very similar (Winters,2008b). The ProtoDravidian terms *baraga and *tena have little if any affinity to the African terms for millet (Winters, 1999, 1999b, 2000).

The Kol term for millet 'sonna', is very similar to the terms for millet used by the Wolof 'suna' (a West Atlantic Language), and Mande 'suna' (a Mande language). The agreement of these terms in sound structure suggest that these terms may be related.

The sound change of the initial /s/ in the African languages , to the /c/ in Tamil and Malayalam is consistent with the cognate Tamil and Malayalam terms compared by Aranavan(1979 ,1980;) and Winters (2008b). Moreover, the difference in the Kol term ' soona',which does retain the complete African form indicates that the development in Tamil and Malayalam of c < s, was a natural evolutionary development in some South Dravidian languages. Moreover, you will also find a similar pattern for other

Malinke and Dravidian cognates, e.g., buy: Malinke 'sa, Tamil cel; and road: Malinke 'sila', Tamil 'caalai'.

The sound change of the initial /s/ in the African languages , to the /c/ in Tamil and Malayalam is consistent with the cognate Tamil and Malayalam terms compared by Aranavan(1979 ,1980;) and Winters (1981, 1994). Moreover, the difference in the Kol term ' soona',which does retain the complete African form indicates that the development in Tamil and Malayalam of c < s, was a natural evolutionary development in some South Dravidian languages. Moreover, you will also find a similar pattern for other Malinke and Dravidian cognates, e.g., buy: Malinke 'sa, Tamil cel; and road: Malinke 'sila', Tamil 'caalai'.

Ceramic Traditions

The Kushites used three types of pottery 1) Wavy Line , 2) Dotted Wavy Line and 3) Red-and-Black pottery (BRW). Pottery types 1 and 2 are analogous to European Bell Beaker and Corded ware.

The BRW industry diffused from Nubia, across West Asia into Rajastan, and thence to East Central and South India. Singh (1992) made it clear that he believes that the BRW radiated from Nubia through Mesopotamia and Iran downward into India.

BRW is found at the lowest levels of Harappa and Lothal dating to 2400BC. T.B. Nayar (1977) proved that the BRW of Harappa has affinities to predynastic Egyptian and West Asian pottery dating to the same time period.
After 1700 BC, with the end of the Harappan civilization BRW spread southward into the Chalcolithic culture of

Malwa and Central India down to Northern Deccan and eastward into the Gangetic Basin.

As the Sahara and Sahel became more arid the Kushites began to migrate into Eurasia and Europe. Using boats the Kushites moved down ancient waterways many now dried up, to established new towns in Asia and Europe after 3500 BC.

Discussion

According to Sergent (1992), the Dravidian populations are not autochthonous to India , they are of African origin. The archaeological evidence also supports an African origin for the Dravidian speaking people (Lal, 1963; Winters, 2007,2008).

Researchers have conclusively proven that the Dravidians are related to the Niger-Congo speaking group and they originally lived in Nubia (Lal,1963). The Dravidians and C-Group people of Nubia used 1) a common BRW (Lal,1963); 2) a common burial complex incorporating megaliths and circular rock enclosures (Lal,1963); and 3) a common type of rock cut sepulcher (Lal,1963) and writing system (Winters, 2007,2008).

The linguistic and anthropological data make it clear that the Dravidian speaking people were part of the C-Group people who formed the backbone of the Niger-Congo speakers. It indicates that the Dravidians took their red-and-black pottery with them from Africa to India, and the cultivation of millet. The evidence makes it clear that the genetic evidence indicating a Holocene migration to India for the

Dravidian speaking people is wrong. The Dravidian people given the evidence for the first cultivation of millet and red-and-black pottery is firmly dated and put these cultural elements in the Neolithic. The evidence makes it clear that genetic evidence cannot be used to effectively document historic population movements.

The Dravidian and Mande speakers began to migrate out of Africa by 3000BC. They were part of the C-Group (Lal, 1963, Winters, 2014). They first settled in Iran and from here expanded into Central Asia Europe and the Indus Valley (Winters,2018).

The Niger-Congo speakers or Kushites formerly lived in the highland regions of the Fezzan and Hoggar until after 4000 BC. The ancestors of the Kushites were the Ounanians who spread the Ounan-Harfian toolkit, pottery and arrows from throughout North Africa, into Iberia and the Levant. Originally huntergatherers the Proto-Niger-Congo people developed an agro-pastoral economy which included the cultivation of millet, and domestication of cattle (and sheep). It was these Kushites who introduced mtDNA U6, M1, T2, X and K; and Y-Chromosome R1b into Eurasian from their African homeland in the Sahel-Sahara.

The Hattic speaking people were members of the Kushite tribe called Tehenu. They were probably called Hati (pl. Hatiu), by the Egyptians. The Hattic name for themselves: Kashka , ḫ3st or Kushite Nation. The Hattic people probably carried Rb1.

The language of the Hittites was more than likely a lingua franca, with Hattic, at its base. In Western Anatolia many languages were spoken including Hattic, Palaic, Luwian and Hurrian, the nationalities there used Nesa as a lingua franca. For example, the king of Arzawa, asked the Egyptians in the Amarna Letters, to write them back in Nesumnili rather than Egyptian (Singer, 1981).

Steiner (1981) notes that "In the complex linguistic situation of Central Anatolia in the 2nd Millennium B.C., with at least three, but probably more different
languages being spoken within the same area there must have been the need for a language of communication or lingua franca whenever commercial transaction or political enterprises were undertaken on a larger scale" (Singer, 1981) .

This led Steiner (1981) to conclude that "moreover the structure of Hittite easily allowed one to integrate not only proper names, but also nouns of other languages into the morphological system. Indeed, it is a well known fact the vocabulary of Hittite is strongly interspersed with lexemes from other languages, which is a phenomenon typical of a "lingua franca". This supports the view that the Anatolians were not predominately Indo-European speakers. The relationship between the Dravidian languages and the Kushite languages spoken in Anatolia explains the presence of the Y-Chromosome R1a clade in this area in ancient times.

Hattians lived in Anatolia. They worshipped Kasku and Kusuh. They were especially prominent in the Pontic mountains. Their sister nation in the Halys Basin were the Kaska tribes. The Kaska and Hattians share the same names for gods, along with personal and place-names. The Kaska had a strong empire which was never defeated by the Hittites.

The Kushites spread cultivation of Pennisetum millet and cattle herding into Anatolia, South Asia and Europe. As cattle herding Kushites frequently moved from place to place millet was an ideal domesticate.

Millet was an especially favorite crop for the mobile Kushites because the grains are 1) a high yield per plant; 2) millet is drought tolerant and can be grown in various terrains; 3) millet has a short growing season so pastoralists could grow and harvest their crops in time to move their camp(s); and 4) the panicum millet has shallow roots so Kushite farmers could cultivate the crop with a hoe (Miller et al., 2010).

Ounanians crossed the Straits of Gibraltar and settled Iberia. Here they met Iberian hunter-gatherers. Between 3200-2900 BC, African culture and people began to migrate into Iberia and introduced megaliths and the Bell Beaker culture (Lahovary, 1963). Spanish researchers accepted the reality that the Iberia Peninsula owed the major parts of Neolithic Iberia to African immigrants (Lahovary, 1963; MacWhite.1947; Winters, 2017b).
MacWhite (1947) and Olalde et al., (2017) claims there was a close relationship between Iberia and Britain. These researchers admit that Portugal and Brittany were settled by Megalithic Africans who founded respectively the Mugem and Teviec sepultures (MacWhite, 1947).
Iñigo Olalde et al., (1917) discuss the spread of Bell Beaker culture across Europe 2.7 kya. These researchers found limited genetic affinity between individuals from Iberia and central Europeans. Iñigo Olalde et al., (2017) concludes that migration probably played an insignificant mechanism

in the spread of R1 within the two areas.

The African Sahara and Morocco was a major source for the **Bell Beaker** and **Corded Ware** cultural complex. The Proto-Beaker pottery dates back to 4500 BC in the Sahara . Daugas et al., (1989) provides a number of radio carbon dates for the Bell Beaker complex in North Africa. We find Beaker Bell ware dating to 3700 BC in Morocco. By 2700 BC we see the expansion of Beaker complex into Iberia (Daugas,1989). The Iberian Bell Beaker complex is associated with the *"Maritime tradition"* (Mathieson et al2015, 2017; Turek,2012).

There are numerous Bell Beaker sites in the Sahara and Morocco. A center of the Moroccan Beaker complex ceramics and arrowheads come from Hassi Ouenzga and in the cave of Ifri Ouberrid . Artifacts found at these sites are similar to Iberian Beaker complex forms (Mikdad, 1998).

The interesting fact about the discovery of these artifacts is that they were widespread across the Middle Atlas mountains at sites such as El-Kiffen, Skhirat – de Rouazi, Kehf, That el Gher and Ifri Ouberrid (Guilaine, 1976;Mikdad, 1998; Nekkal and Mikdad, 2014). This finding matches Turek (2012); which explains the spread of typically beaker style stamped decoration Bell Beaker culture pottery from Morocco into Iberia, and thence the rest of Europe.

Toomas Kivisild and Mathieson et al., (2017) , provides a detailed discussion of R1 in prehistoric Europe. One of the most interesting finding was the presence of V88 in ancient Europe (Kivisild, 2017;Mathieson,2015; Olalde et al,2017). It is also interesting to note that the European Agro-Pastoral populations associated with Bell Beaker and Yamnaya carry the genomes associated with Africans recorded in Table 1.

MANUFACTURED GENETIC ORIGIN

Table 1: Shared African and Eurasian R1b Clades

Nations and Populations	R1b M345	R1b1 L-278	R1b1a R-L754	R1b1a1a2	R1b1a2	R1b1b2	R1b1b2a1	Source
Akele		0.02						Berniell-Lee et al, 2009
Baka		0.03						Berniell-Lee et al, 2009
Bakola		0.045						Berniell-Lee et al, 2009
Benga		0.041						Berniell-Lee et al, 2009
Boni	4.8							Hirbo, 2011
Burji	4.3							Hirbo, 2011
Cabinda	9.5							Hirbo, 2011
Dama		0.21						Berniell-Lee et al, 2009
Fang		0.2						Berniell-Lee et al, 2009
Fali					20			Haber et al, 2016
Fulbe					11.1			Haber et al, 2016
Fante	6							Hirbo, 2011
Equatorial Guinea						0.096		Gonzalez et al, 2013
Gabon			0.086					Gonzalez et al, 2013
Guinea-Bissau						2.2		Carvalho et al, 2011
Hausa	40.63							Hirbo, 2011
Herero	16				8			Hirbo, 201 ; Haber et al, 2016
Kola		0.056						Berniell-Lee et al, 2009
Kanuri	36.7				14.3			Hirbo, 2011 ; Haber et al, 2016
Laal				23				
Mada					82.4			Haber et al, 2016
Mafa					87.5			Haber et al, 2016
Mandara	37.5							Hirbo, 2011
Moundang					66.78			Haber et al, 2016
N'Djamama				15				
Ndumu	0.11							Berniell-Lee et al, 2009
Ngambi					9.1			Haber et al, 2016
Nzebi		0.035						Berniell-Lee et al, 2009
Obamba		0.11						Berniell-Lee et al, 2009
Orungu		0.045						Berniell-Lee et al, 2009
Ouideme					95.5			Haber et al, 2016
Punu		0.12						Berniell-Lee et al, 2009
Rangi	3.1							Hirbo et al, 2011
Shake		0.023						Berniell-Lee et al, 2009
Sara				20				
Tali					9.1			Haber et al, 2016
Teke		0.104						Berniell-Lee et al, 2009
Yoruba					4.8			
Khoisan						2.2		Wood et al, 2005
Khoisan						6		Hirbo et al, 2011
Khomani San							10	Henn et al, 2011

Europe								
Individual	Country	Haplogroup						
14916	Serbia	R1b1a						Mathieson et al, 2017
14081	Romania	R1b1a						Mathieson et al, 2017
15235	Serbia	R1b1a						Mathieson et al, 2017
15237	Serbia	R1b1a						Mathieson et al, 2017
15240	Serbia	R1b1a						Mathieson et al, 2017
SS772. E1.L1	Serbia	R1b1a						Mathieson et al, 2017
15232	Serbia	R1b1a						Mathieson et al, 2017
15232	Serbia	R1b1a						Mathieson et al, 2017
14432	Latvia	R1b1a1a						Mathieson et al, 2017
14434	Latvia	R1b1a1a						Mathieson et al, 2017
14439	Latvia	R1b1a1a						Mathieson et al, 2017
14626	Latvia	R1b1a1a						Mathieson et al, 2017
14628	Latvia	R1b1a1a						Mathieson et al, 2017
14630	Latvia	R1b1a1a						Mathieson et al, 2017
14436	Latvia	R1b1a1a						Mathieson et al, 2017
14627	Latvia	R1b1a1a						Mathieson et al, 2017
10122	Samara	R1b1a						Mathieson et al, 2017
I1734	Ukraine	R1b1a						Mathieson et al, 2017
I3718	Ukraine	R1b1a						Mathieson et al, 2017
I4114	Ukraine	R1b1a						Mathieson et al, 2017
SS883.E1.L1	Ukraine	R1b1a						Mathieson et al, 2017
SS890.E1.L1	Ukraine	R1b1a						Mathieson et al, 2017
SS892.E1.L1	Ukraine	R1b1a						Mathieson et al, 2017
SS893.E1.L1	Ukraine	R1b1						Mathieson et al, 2017
Vilabruna	Italy	R1b1a						

Chuan et al (2019), recently published an article that discussed the genomes in Neolithic Europe. The researchers noted that "The steppe groups from Yamnaya and subsequent pastoralist cultures show evidence for previously undetected farmer-related ancestry from different contact zones, while Steppe Maykop individuals harbour additional Upper Palaeolithic Siberian and Native American related ancestry. " The researchers added that " [it] is also visible in the Y-chromosome haplogroup distribution, with R1/R1b1 and Q1a2 types in the Steppe and L, J, and G2 types in the Caucasus cluster " This is interesting because the Kushites of the Yamnaya culture mainly carried R1b. These Kushites practiced an agro-pastoral culture.

The Yamnaya culture bearers came from the Levant. These people based on ancient Bullae identified themselves as Kushites. The Levant had early been settled by people from the Nile Valley. Beginning as early as 5000 years ago Kushites the **ḥ3št**, lived from the Nile Valley below Egypt, all the way to the Levant and Anatolia. The Kushites belonged to the C-Group culture of Nubia. The Kushites spoke Niger-Congo and Dravidian languages .

The Niger-Congo (NC) Superfamily of languages is the largest family of languages spoken in Africa. Researchers have assumed that the NC speakers originated in West Africa in the Inland Niger Delta. The research indicates that the NC speakers originated in the Saharan Highlands 12kya and belonged to the Ounanian culture .

The Kushites were called **ḥ3st** in Africa and the Levant. Kushites had early settled in the Levant since Narmer times. We find Narmer's name on jars and serekhs from excavations in Israel and Palestine , for example Tel Erani, Arad, 'En Besor, Halif Terrace/Nahal Tillah and more(4). A bulla dating to this period makes it clear that this part of the Negev was called *ḥ3ts.t* ("Kush") or *ḥ3s.tj* ("Kushite").

Chuan et al (2019) made it clear that, " Recent ancient DNA studies have resolved several longstanding questions regarding cultural and population transformations in prehistory. One important feature is a cline of European hunter-gatherer (HG) ancestry that runs roughly from West to East (hence WHG and EHG; blue component in Fig. 2a,c). This ancestry differs from that of Early European farmers, who are more closely related to farmers of northwest Anatolia21,22 and also to pre-farming Levantine individuals9. The near East and Anatolia have long-been seen as the regions from which European farming and animal husbandry emerged. In the Mesolithic and Early Neolithic, these regions harboured three divergent populations, with Anatolian and Levantine ancestry in the west, and a group with a distinct ancestry in the east. The latter was first described in Upper Pleistocene individuals from Georgia (Caucasus hunter-gatherers; CHG)13 and then in Mesolithic and Neolithic individuals from Iran9,23. The following millennia, spanning the Neolithic to BA, saw admixture between these ancestral groups, leading to a pattern of genetic homogenization of the source populations9. North of the Caucasus, Eneolithic and BA individuals from the Samara region (5200–4000 BCE) carry an equal mixture of EHG- and CHG/Iranian ancestry, so-called 'steppe ancestry' 13 that eventually spread further west18,19, where it contributed substantially to present-day Europeans, and east to the Altai region as well as to South

Asia9 "

This makes it clear that there was continuity between the Black populations that spread the contemporary DNA into Eurasia. It was Africans from the Nile Valley and North Africa that invented the ceramic assemblages that are called the Bell Beaker and Corded Ware in Eurasia.

This makes it clear that the V88 sub-clades R-L278 and R-L754. , had relatives in Early Neolithic samples from across a wide geographic area from Iberia, Germany to Samara (Kivisild, 2017; Olalde et al,2017; Winters,2017b). This would place carriers of relatives of V88 among the Yamnaya and Bell Beaker people. Given the wide distribution of M269 in Africa, the carriers of this haplogroup in Neolithic Europe were probably also Africans since the Bell Beaker people/culture originated in Morocco as noted by Turek (2012).

CONCLUSION

In conclusion, Millet cultivation , and herding cattle was spread across Europe by Kushite Niger-Congo and Dravidian speakers this explains the spread of R1a and R1b by the agro-pastoral populations associated with the late Yamnaya periods. Sub-Saharan Africans had a long history in the Levant and Europe prior to the raise of the Caucasus hunter gathers (CHG) and European farmers (EF) populations(Holliday,2000; Jones et al, 2015) . In Table 1, we provide a comparison of Eurasians and African genomes (Coia et al, 2005;Cruciani et al, 2010; Haak et al., 2015; Haber et al,2016; Henn et al., 2011; Kivisild, 2017; Mathieson et 2015, 2017 ;Olalde et al,2017; Winters, 2016; Wood et al, 2005).

The second migration of Africans came from the East. The people from the Steppes were dark skinned as illustrated by Cassidy et al (2020) in Table 12.

	Fine Structure Cluster	Sample	Skin Predict	Hair Predict	Eye Predict	
2	4. Northwestern HG	KGH6	Dark Black	Dark Brown	Blue?	0
3	4. Northwestern HG	SRA62 (High Coverage)	Dark Black	Dark Brown	Blue	0
4	4. Northwestern HG	SRA62	Dark	Dark Brown	Blue	0
5	4. Northwestern HG	CheddarMan	Dark Black	Brown/Dark Br	Brown?	0
6	4. Northwestern HG	Loschbour (High Coverage)	Intermediate to Dark Black	Dark Brown	Blue	0
7	4. Northwestern HG	Loschbour	Intermediate to Dark Black	Brown/Dark Br	Brown?	0
8	4. Northwestern HG	Bichon	Dark Black		Brown	0
9	4. Northwestern HG	Bichon (High Coverage)	Dark Black		Brown	0
10	5. Spanish HG	Chan	Dark Black		Brown	0
11	5. Spanish HG	LaBrana1	Dark Black	Brown/Dark Br	Brown?	0
12	5. Spanish HG	Canes	Intermediate	Dark Brown	Blue?	0
13	2. Latvian HG	ZVEJ25	Dark Black	Brown/Dark Br	Blue?	0
14	2. Latvian HG	ZVEJ27	Dark Black	Brown/Dark Br	Blue?	0
15	2. Latvian HG	ZVEJ32	Dark Black	Brown/Dark Br	Brown?	0
16	1. Central and South Eastern HG	SC1_Meso	Dark Black	Brown/Dark Br	Brown	0
17	1. Central and South Eastern HG	SC2_Meso	Dark Black		Brown	0
18	1. Central and South Eastern HG	OC	Dark Black	NA	NA	A
19	1. Central and South Eastern HG	KO1	Intermediate to Dark	Brown/Dark Br	Blue?	0
20	3. Eneolithic Romania	G8	Pale to Intermediate	Brown/Dark Br	NA	A
21	5. Eastern HG	Karelia	Intermediate to Dark	Dark Brown	Brown	

	A Fine Structure Cluster	B Sample	C Skin Predict	D Hair Predict	E Eye Predict	
161	8. Northern Iron	NO3423_AngSax	Pale to Intermediate	Blond/Lightest H		
162	12. Steppe Late Bronze Age onwards	Russia_IA1	Intermediate to Dark	Brown/Dark Br	Brown	
163	12. Steppe Late Bronze Age onwards	Russia_IA2	Intermediate	Dark Brown	Brown	
164	12. Steppe Late Bronze Age onwards	Russia_IA3	Dark to Dark Black		Brown	
165	12. Steppe Late Bronze Age onwards	Karasuk_Sab1	Pale to Intermediate	Brown/Dark Br	Brown	
166	12. Steppe Late Bronze Age onwards	Karasuk_Sab1 (High Cover	Intermediate to Dark	Brown/Dark Br	Brown	
167	12. Steppe Late Bronze Age onwards	Karasuk_Arb1 (High Covera	Intermediate to Dark		Brown	
168	12. Steppe Late Bronze Age onwards	Karasuk_Arb1	Intermediate to Dark	Brown/Dark Br	Brown	
169	12. Steppe Late Bronze Age onwards	Karasuk_Arb2	Intermediate to Dark	Dark Brown	Brown	
170	12. Steppe Late Bronze Age onwards	Karasuk_Arb3	Dark			
171	12. Steppe Late Bronze Age onwards	Karasuk_By1	Intermediate to Dark	Dark Brown	Brown?	
172	12. Steppe Late Bronze Age onwards	Karasuk_By3	Pale to Intermediate	Dark Brown	NA	
173	13. Steppe Copper and Bronze Age	Afanasievo_Bat1	Dark	Dark Brown	Brown	
174	13. Steppe Copper and Bronze Age	Yamnaya_Kal1	Intermediate to Dark		Brown	
175	13. Steppe Copper and Bronze Age	Yamnaya_Kal2	Intermediate to Dark		Brown	
176	13. Steppe Copper and Bronze Age	Andronovo_Kyt2	Intermediate to Dark	Brown/Dark Br	Brown	
177	13. Steppe Copper and Bronze Age	Andronovo_Kyt1	Pale to Intermediate	Lightest Brown	Blue	
178	13. Steppe Copper and Bronze Age	Mezhovskaya_Kap1	Dark	Lighter Brown	NA	
179	13. Steppe Copper and Bronze Age	Sintashta_Bol1	Pale to Intermediate	Blond/Lightest	Brown	

Table 12,Cassidy, L.M., Maoldúin, R.Ó., Kador, T. et al. A dynastic elite in monumental Neolithic society. Nature 582, 384–388 (2020). https://doi.org/10.1038/s41586-020-2378-6

The Niger-Congo speaking Africans and Dravidian speakers carried R1b and R1a first into Anatolia (Winters,

2010,2017b), and thence Europe. We can see from Table 1, that Africans carry R1b1 which is associated with the CHG, and R1b1a (Winters, 2017b) is related to the EF. Up until 2010, the R1b1a clade, was recognized as V88 (Cruciani et al,2010).

As a result, the "Eurasian" admixture, found among the West Africans in East, Central, West and South Africa is in reality African genomes passed onto the Eurasians when the Kushites migrated into Eurasia 4kya from Africa (Winters, 2010,2010b, 2017b). Other "Eurasian" genomes of African origin were deposited in Eurasia first by African hunter-gatherers in Iberia carrying Y-Chromosome R1 clades eastward into Eastern Europe and the Steppes (Winters,2017b), as indicated by the aDNA found in Vilabruna man (R1b1a) and Samara (R1b1) that are relatives of V88 (Kivisild,2017; Winters, 2017b).

Kivisild (2017) maintains that the CHG R1 clades: R1b1 and R1b1a are distant relatives to V88. In Table 1, we compare African and Eurasian R1 lineages (Coia et al., 2005; Cruciani et al., 2010; Haak et al., 2015; Haber et al,2016; Henn et al., 2011; Kivisild, 2017; Mathieson et 2015, 2017 ;Olalde et al,2017; Winters, 2016; Wood et al., 2005). As illustrated in Table 1, the global nature of R-L278 and R-L754 across Africa, east to west and north to south among populations that fail to carry the R1 clades, but lack Neanderthal ancestry does not support the conclusions of Haber et al., (2016) that there was ever a backflow from Eurasia into Chad of Eurasian ancestry, once, let alone, twice in the past 10ky.

The findings of Haber et al., (2016) are unfounded and cannot be confirmed because, there is no archaeological evidence that Eurasians made their way back to West and

Central Africa. Lacking archaeological evidence of Eurasians in West and Central Africa, the so-called Eurasian admixture among the varied African populations carrying the R1 lineage reflects the fact that the Eurasian YChromosome R1, is really of West and Central African origin, not Eurasian.

The archaeological evidence does make it clear there were two migrations into Western Eurasia by Africans. The first migration to Eurasia from Africa, was by foragers carrying R1b1 and R1b1a, who first settled Iberia and Italy and migrated eastward with millet.

 These Africans were Kushites who first settled in Anatolia and spread eastward into India, with millet, and into the Steppe region, and from there westward into Iberia, and eventually Britain. The Kushites from Africa that settled Eurasia, were Niger-Congo and Nilo-Saharan speakers who not only carried R1b1 and R1b1a, but also R1-M269. The R1-M269 clade has the highest frequency among the early European Agro-pastoralists.

This archaeological evidence further indicates that the Bell Beaker and Corded ware cultures (were probably descendant from the Wavy line and Dotted pattern pottery of the African Neolithic) and Bell Beaker pottery from Morocco and the Saharan region spread across Europe with the EF agro-pastoral civilization.

Finally, Cattle herding and millet cultivation were all introduced to Eurasia by African and Dravidian speakers. Since, these culture elements are all associated with the Niger-Congo- Dravidian and NiloSaharan carriers of R1a and R1b, these Y-Chromosome clades were introduced to

Eurasia by Africans. This is the only way we can explain the "dilution" of Neanderthal DNA in the Near East, as maintained by Haber et al., (2016).

REFERENCES

Anselin A (1982). Le Mythe d'Europe de l'Indus à la Crète.

Anselin A (1989). Le Lecon Dravidienne",Carbet Revue Martinique de Sciences Humaines, no.9:7-58

Aravanan K P. Notable negroid elements in Dravidian India, Journal of Tamil Studies, 1980, pp.20-45.

Aravanan K P (1976). Physical and cultural similarities between Dravidians and Africans. Journal of Tamil Studies 10 23-27.

Aravanan K P (1979). Dravidians and Africans, Madras.

Barnabas S, Shouche Y and Suresh CG (2005). High resolution mtDNA studies of the Indian population: Implications for Paleolithic settlement of the Indian Subconinent, Annals of Human Genetics, 1-17.

Berniell-Lee G, Calafell F, Bosch E, Heyer E, Sica L, Mouguiama-Daouda P, van der Veen L, Hombert J-M, Quintana-Murci L and Comas D (2009). Genetic and Demographic Implications of the Bantu Expansion: Insights from Human Paternal Lineages. Molecular Biology and Evolution 26(7) 1581- 1589 doi:10.1093/molbev/msp069

Bork, Ferdinand (1909). Die Mitanni Sprache, Mitteilungen der Vorderasiatische Gesellschaft, Parts ! and 2.

Brown, George William. (1930) The Possibility of a Connection between Mitanni and the Dravidian Languages , Journal of the American Oriental Society, Vol. 50 (1930), pp. 273-305

Gupta, A. How old is the Rig Veda {Part 2}. Retrieved: 14 January 2004

http://www.sawf.org/newedit/edi40205200/musings.asp
Carvalho M. , Brito P, Bento AM et al., (2011). Paternal and maternal lineages in Guinea-Bissau population. Forensic Science International Genetics , 5: 114–116.

Chuan-Chao Wang et al. (2019). Ancient human genome-wide data from a 3000-year interval in the Caucasus corresponds with eco-geographic regions. NATURE COMMUNICATIONS (2019) 10:590 https://doi.org/10.1038/s41467-018-08220-8

Coia V, Destro-Bisol G, Verginelli F, Battaggia C, Boschi I, Cruciani F, Spedini G, Comas D and Calafell F (2005). Brief communication: mtDNA variation in North Cameroon: lack of Asian lineages and implications for back migration from Asia to sub-Saharan Africa. American Journal of Physical Anthropology 128(3) 678-81. Available: http://www3.interscience.wiley.com/cgibin/fulltext/11049 5269/PDFSTART

Cruciani F, Santolamazza P, Shen P, Macaulay V, Moral P and Olckers A (2002). A Back Migration from Asia to Sub-Saharan Africa is supported by High-Resolution Analysis of Human Y-chromosome Haplotypes. American Journal of Human Genetics 70 1197-1214.

Cruciani F, Trombetta B, Sellitto D, Massaia A, Destro-Bisol G, Watson E, Beraud Colomb E, Dugoujon JM, Moral P and Scozzari R (2010). Human Y chromosome haplogroup R-V88: a paternal genetic record of early mid Holocene trans-Saharan connections and the spread of Chadic languages. European Journal of Human Genetics 18 800–807. doi:10.1038/ejhg.2009.231

Daugas JP, Raynal J-P, Ballouche A, Occhietti S, Pichet P, Evin J, Texier J-P and Debenath A. (1989). Le Neolithique Nord-Atlantique Du Maroc: Premier Essai De Chronolgie Par Le Radiocarbon, (C.R. Academy of Sciences, Paris, 308,

France), Serie II, 681-687.

Domínguez EF (2005). Polimorfismos de DNA mitocondrial en poblaciones antiguas de la Cuenca mediterránea. PhD Thesis, Universitat de Barcelona, Departament de Biologia Animal.

Drake N A, Roger M. Blench, Simon J. Armitage, Charlie S. Bristow, and Kevin H. White. (2010). Ancient watercourses and biogeography of the Sahara explain the peopling of the desert, Proceedings of the National Academy of Scienece. 2011 108 (2) 458-462; published ahead of print December27, doi:10.1073/pnas.1012231108

Diakonoff, I M and PL Kohl (1990). Early Antiquity. Chicago: University of Chicago Press.

El Mosallamy AHS (1986). Libyco-Berber relations with ancient Egypt: The Tehenu in Egyptian records. In L. Borchardt, Das Grabdenkmal des Konigs Sahure 51-68, 2, Table 1.

Fregel R, et al., (2017). Neolithization of North Africa involved the migration of people from both the Levant and Europe. bioRxiv 191569; doi: https://doi.org/10.1101/191569

Gonzalez et al., (2012). The genetic landscape of Equatorial Guinea and the origin and migration routesof the Y chromosome haplogroup R-V88. European Journal of Human Genetics 21(3)324-331. Available: doi: 10.1038/ejhg.2012.167.

Guilaine J (1976). La Civilisation Du Vase Campaniforme Dans Le Midi De La France, Actes Du Syposium Sur La Civilisation Des Gobelets Campaniformes, (Germany, Oberried, Bussum-Haarlem) 351-370.

Gupta, A (No Date). How old is the Rig Veda (Part2). Retrieved: 14 January 2004 http://www.sawf.org/newedit/edi40205200/musings.asp

Haak W, Lazaridis I, Patterson N, Rohland N, Mallick S,

Llamas B, Brandt G et al., (2015). Massive migration from the steppe was a source for Indo-European languages in Europe. Nature 522 207–211. doi:10.1038/nature14317

Haber M., Massimo Mezzavilla, Anders Bergström, Javier Prado-Martinez, Pille Hallast, Riyadh Saif-Ali et al., (2016). Chad Genetic Diversity Reveals an African History Marked by Multiple Holocene Eurasian Migrations, The American Journal of Human Genetics, Volume 99, Issue 6, 1316-1324, ISSN 0002-9297, http://dx.doi.org/10.1016/j.ajhg.2016.10.012

Hernández CL, Soares P, Dugoujon JM, Novelletto A et al., (2015). Early Holocenic and Historic mtDNA African Signatures in the Iberian Peninsula: The Andalusian Region as a Paradigm. PLoS ONE 10(10). Available: https://www.ncbi.nlm.nih.gov/pubmed/26509580

Holiday T (2000). Evolution at the Crossroads: Modern Human Emergence in Western Asia. American Anthropologist 102(1) 54-68.

Homburger L. 1948. Elements Dravidiens en peul. Journal Societe de Africa, 18(2): 135-143.

Homburger L 1957. Les Langues Negro-Africaines et les peoples qui les parlent. Paris: Payot.

Hunt H.V et al., (2008) Millets across Eurasia: chronology and context of early records of the genera Panicum and Setaria from archaeological sites in the Old World, Vegetation History and Archaeobotany, vol. 17, Suppl 1, S5–S18.

Jelinek J (1985). "Tillizahren,the Key Site of the Fezzanese Rock Art". Anthropologie (Brno), 23(3):223-275.

Jones ER, Gonzalez-Fortes G, Connell S, Siska V, Eriksson A, Martiniano R and Bradley DG (2015). Upper Palaeolithic genomes reveal deep roots of modern Eurasians. Nature Communications 6 8912. http://doi.org/10.1038/ncomms9912

Kefi R., Meriem Hechmi, Chokri Naouali, Haifa Jmel, Sana Hsouna, Eric Bouzaid, Sonia Abdelhak, Eliane Beraud-Colomb & Alain Stevanovitch (2016). On the origin of Iberomaurusians: new data based on ancient mitochondrial DNA and phylogenetic analysis of Afalou and Taforalt populations, Mitochondrial DNA Part A, 29:1, 147-157, DOI: 10.1080/24701394.2016.1258406

Kılınç, Gülşah Merve et al., (2016). The Demographic Development of the First Farmers in Anatolia Current Biology , Volume 26 , Issue 19 , 2659 – 2666

Kivisild, Toomas, Katrin Kaldman, Mait Metspalu, Juri parik, Surinder Papiha (1999).The Place of the Indian mtDNA Varients in the Global Network of Maternal Lineages and the Peopling of the OldWorld. In Genomic Diversity, (Ed.) R. Papiha Deka (pp.135-152). S.S. Kluwer/Plenum Publishers. http://evolutsioon.ut.ee/publications/Kivisild1999b.pdf

Kivisild T (2017). The study of human Y chromosome variation through ancient DNA. Human Genetics 136(5) 529–546.

Lahovary N (1963). Dravidian Origins and the West. India, Madras: Longmans.

Lal B B (1963) "The Only Asian expedition in threatened Nubia:Work by an Indian Mission at Afyeh and Tumas", The Illustrated London Times , 20 April.

Lazaridis I, Nadel D, Rollefson G, Merrett DC et al., (2015). The genetic structure of the world's first farmers. bioRxiv doi: https://doi.org/10.1101/059311

Le Quellec J-L (2011). Arcs et bracelets d'archers au Sahara et en Égypte, Avec Une Nouvelle Proposition De Lecture Des "Nasses" Sahariennes. (CEMAf - Centre d'Etudes des Mondes Africains, Johannesburg, Africa). Available: https://halshs.archives-ouvertes.fr/halshs-00696540/document

Levy T E, David AlonYorke M. RowanYorke M. Rowan (1997). Egyptian-Canaanite Interaction at Nahal Tillah, Israel (ca. 4500-3000 B. C. E.): An Interim Report on the 1994-1995 Excavations.

MacWhite E (1947). Studios sobre las relaciones atlanticas de la peninsula hispanica en la edad del bronce. Dissertationes Matritenses 12.

Metspalu M (2005). Through the course of prehistory in India: Tracing the mtDNA Trail. Dissertation Biologicae Universitatis Tartnensis 114, Tartu University Press.

Metspalu , Mait, Toomas Kivisild, Ene Metspalu, Jüri Parik, Georgi Hudjashov , Katrin Kaldma, Piia Serk, Monika Karmin, Doron M Behar, et al. (2004).Most of the extant mtDNA boundaries in South and Southwest Asia were likely shaped during the initial settlement of Eurasia by anatomically modern humans. BMC Genetics 2004, 5:26. http://www.biomedcentral.com/1471-2156/5/26

Mathieson I, Lazaridis I, Rohland N et al., (2015). Genome-wide patterns of selection in 230 ancient Eurasians. Nature 528 499–503. doi:10.1038/nature16152

Mathieson I, Roodenberg SA, Posth C et al., (2017). The Genomic History of Southeastern Europe. Available: http://biorxiv.org/content/early/2017/05/09/135616

McIntosh, S.K. and McIntosh R J (1983). "Forgotten Tells of Mali". Expedition, 38 .

Mitchell P , Paul Lane (Ed.),(2013). The Oxford Handbook of African Archaeology. Oxford .

Martinez L, et al., (2007). Paleolithic Y-haplogroup heritage predominates in a Cretan highland plateau, European Journal of Human Genetics. http://www.nature.com/ejhg/journal/v15/n4/full/52017 69a.html

Mikdad A (1998). Étude Préliminaire Et Datation De Quelques Éléments Campaniformes Du Site De Kehf-el-

Baroud, Maroc, (AVA-Forschungen, Bd. 18, Mainz, Germany) 243-252.

Miller N F , Robert N Spengler, Michael Frachetti (2010). Millet cultivation across Eurasia: Origins, spread, and the influence of seasonal climate, The Holocene , Vol. 26 10:1566-1575

Motuzaite-Matuzeviciut G et al., (2013). The early chronology of broomcorn millet (Panicum miliaceum) in Europe. Antiquity 87(338) 1073–1085.

Müller J and van Willigen S (2001). New radiocarbon evidence for European Bell Beakers and theconsequences for the diffusion of the Bell Beaker phenomenon. In Bell Beakers Today: Pottery, People, Culture, Symbols in Prehistoric Europe. Proceedings of the International colloquium, Riva del Garda Trento, Italy, (edition Nicolis, F., 59–80).

N'Diaye CT (1972) The relationship between Dravidian languages and Wolof. Annamalai University Ph.D. Thesis.

Nayar TB (1977). The Problem of Dravidian Origins.

Nekkal F and Mikdad A (2014). Quelques données sur la découverte de céramiques campaniformes au Maroc [Some data on the discovery of Bell Beaker pottery in Morocco]. International Journal of Innovation and Applied Studies 8(2) 632-638.

Olalde I, Brace S, Allentoft ME, Armit I, Kristiansen K et al., (2017). The Beaker Phenomenon and the Genomic Transformation of Northwest Europe. bioRxiv vol doi: https://doi.org/10.1101/135962

Pickrel J Kl, Nick Patterson, Po-Ru Loh, Mark Lipson, Bonnie Berger, Mark Stoneking, Brigitte Pakendorf, and David Reich (2014). Ancient west Eurasian ancestry in southern and eastern Africa, Proceedings of the National Academy of Science, 111 (7) 2632-2637; published ahead of print February 3, 2014, doi:10.1073/pnas.1313787111

Potts T (1995). Mesopotamia and the East. Oxford Unversity Committee for Archaeology. Monograph 37.

Prieto-Martínez MP (2011). Perceiving changes in the third millennium BC in Europe through pottery: Galicia, Brittany and Denmark as examples. In Becoming European: The Transformation of Third Millennium Northern and Western Europe (edition Prescott, C. & Glorstad, H., 30–47, (UK, Oxford: Oxbow Books).

Ramana G V, Su B, Jin L, Singh L, Wang, N., Underhill, P. & Chakraborty, R. (2001) Ychromosome SNP haplotypes suggest evidence of gene flow among caste, tribe, and the migrant Siddi populations of Andhra Pradesh, South India. Eur J Hum Genet 9, 695 – 700. http://archive.is/UlNyk

Rosa A, Ornelas C, Jobling MA, Brehm A and Villems R (2007). Y-chromosomal diversity in thepopulation of Guinea-Bissau: A multiethnic perspective. BMC Evolutionary Biology 7 124.

Sagy,H W F (1995) Peoples of the Past: Babylonians. Norman: University of Oklahoma Press, 1995.

Scheinfeldt LB, Soi S and Tishkoff SF (2010). Working toward a synthesis of archaeological, linguistic, and genetic data for inferring African population history. Proceeding National Academy of Science USA 107(Supplement 2) 8931–8938.

Sergent, Bernard (1992). Genèse de L'Inde. Paris: Payot .

Singh H N (1982). History and archaeology of Blackand Red ware. Vedic Books: Manchester

Sharma S, Rai E, Sharma P, Jena M, Singh S, Darvishi K, Bhat AK, Bhanwer AJS, Tiwari PK & Bamezai NK (2009). The Indian origin of paternal R1a1* substantiates the autochthonous origin of Brahmins and the caste system. J of Hum Genet, 54: 47-55.

Sengupta, Sanghamitra, Lev A. Zhivotovsky, Roy King, S. Q.Mehdi,Christopher A. Edmonds,Cheryl-Emiliane T. et

al., (2006). American Journal of Human Genetics 78 202-221.
http://www.journals.uchicago.edu/AJHG/journal/issues/v78n2/42812/42812.html?erFrom=32142698769629830 94Guest

Schuenemann V. J. , Alexander Peltzer, Beatrix Welte (2017). Ancient Egyptian mummy genomes suggest an increase of Sub-Saharan African ancestry in post-Roman periods. Nature Communications 8: 15694. https://www.nature.com/articles/ncomms15694

Stefania Vai , Stefania Sarno, Martina Lari , Donata Luiselli3, Giorgio Manzi, Marina Gallinaro , Safaa Mataich, Alexander Hübner, Alessandra Modi1, Elena Pilli, MaryAnneTafuri, David Caramelli & Savino di Lernia .(2019). Ancestral mitochondrial N lineage from the Neolithic 'green' Sahara . https://www.nature.com/articles/s41598-019-39802-1.pdf

Singer I (1981). Hittites and Hattians in Anatolia at the beginning of the Second Millennium B.C., Journal of Indo-European Studies, 9 (1-2) (1981), pp.119-149.

Singer I (2007). WHO WERE THE KAŠKA? Journal Phasis, Vol 10, No 1-16 , Retrieved 9/8/2017 at http://phasis.tsu.ge/index.php/phasis/article/view/154/htm

Steiner,G. (1981). The role of the Hittites in ancient Anatolia, Journal of Indo-European Studies, 9 (1-2) (1981), 119-149.

Thangaraj, Kumarasamy, Gyaneshwer Chaubey, Vijay Kumar Singh, Ayyasamy Vanniarajan, Ismail Thanseem, Alla G Reddy and Lalji Singh (2006). In situ origin of deep rooting lineages of mitochondrial Macrohaplogroup 'M' in India. BMC Genomics 7 151. http://www.pubmedcentral.nih.gov/articlerender.fcgi?artid=1534032

Trivedi R, Sahoo S, Singh A, Bindu GH, Banerjee J, Tandon M, Gaikwad S, Rajkumar
Sitalaximi T, Ashma R, Chainy GBN, & Kashyap VK. (2008). Genetic imprints of pleistocene origin of Indian populations: A comprehensive Phylogeographic sketch of Indian Y-Chromosomes. Int J Hum Genet, 8(1-2): 97-118

Turek J (2012). Chapter 8 – Origin of the Bell Beaker phenomenon. The Moroccan connection, In: Fokkens, H. & F. Nicolis (edition) 2012: Background to Beakers. Inquiries into Regional Cultural Backgrounds of the Bell Beaker Complex, (Netherland, Leiden: Sidestone Press). Available:
https://www.academia.edu/1988928/Turek_J._2012_Chapter_8_-

Upadhyaya P & Upadhyaya S P (1979). Les liens entre Kerala et l"Afrique tels qu'ils resosortent des survivances culturelles et linguistiques, Bulletin de L'IFAN, no.1, , pp.100-132.

Upadhyaya P & Upadhyaya S P (1976). Affinites ethno-linguistiques entre Dravidiens et les NegroAfricain, Bull.de L'IFAN, 1127-157.

van de Loosdrecht M., Abdeljalil Bouzouggar, Louise Humphrey, Cosimo Posth, Nick Barton.
(2018). Pleistocene North African genomes link Near Eastern and sub-Saharan African human populations. PUBLISHED ONLINE15 MAR 2018, DOI: 10.1126/science.aar8380

Vernet R, Ott M, Tarrou L, Gallin A, Géoris-Creuseveau J (2007). Excavation of the mound of FA 10 (Banc d'Arguin) and its contribution to the knowledge of the culture paleolithical Foum Arguin, northwestern Sahara (Translated from French) J Afr Archaeol 5:17–46.

Watkins W, Thara R , Mowry B , Zhang Y, Witherspoon D, Tolpinrud W , ... Jorde, L. (2008). Genetic variation in

South Indian castes: evidence from Y-chromosome, mitochondrial, and autosomal polymorphisms. BMC Genetics, 9, 86. http://doi.org/10.1186/1471-2156-9-86

Weber S A (1998). Out of Africa: The initial impact of millets in South Asia. Current Anthropology, (1998) 39(2), 267-274.

Wigboldus J S (1996). Early presence of African millets near the Indian Ocean. In J. Reade, The Indian Ocean (pp.75-86), London: The British Museum, 1996.

Winters C (1980). The genetic unity of Dravidian and African languages and culture",Proceedings of the First International Symposium on Asian Studies (PIISAS) 1979, Hong Kong: Asian Research Service.

Winters C (1985). The Proto-Culture of the Dravidians,Manding and Sumerians, Tamil Civilization, 3 (1), 1-9.

Winters C (1986). The Migration Routes of the Proto-Mande". The Mankind Quarterly, 27(1):77-96.

Winters C (1988.Common African and Dravidian Place Name Elements. Sou As Anth, 9(1): 33-36.

Winters C 1989. Tamil, Sumerian,Manding and the Genetic Model. Int J Dr Ling, 18(1): 67-91.

Winters C (1991). "The Proto-Sahara". In The Dravidian Encyclopaedia (pp.553-556), Trivandrum: International School of Dravidian Linguistics. Volume l.

Winters C (1999). ProtoDravidian terms for cattle. International Journal of Dravidian Linguistics, 28, 91-98.

Winters C (1999b). Proto-Dravidian terms for sheep and goats.PILC Journal of Dravidian Studies, 9 (2), 183-87.

Winters C (2000). Proto-Dravidian agricultural terms. International Journal of Dravidian Linguistics, 30 (1), 23-28.

Winters C (2002). Ancient Afocentric History and the Genetic Model. In Egypt vs Greece, Ed by M.K Asante and A. Mazama,.Pp.121-164.

Winters C (2006).Can Parallel Mutation and neutral genome selection explain Eastern African M1 consensus HVS-1 motifs in Indian M haplogroup . http://www.bioline.org.br/pdf?hg07022

Winters C. 2007. Did the Dravidian Speakers Originate in Africa? BioEssays, 27(5): 497-498.

Winters C (2007b). High Levels of Genetic Divergence across Indian Populations. PloS Genetics. Retrieved 4/8/2008 http://www.plosgenetics

Winters C (2008). Origin and Spread of Dravidian Speakers. http://ferrispages.org/ISAR/krepublishers.pdf

Winters C (2008). African millets taken to India by Dravidians. Annals of Botany, (2008) https://academic.oup.com/aob/article/100/5/903/13606 0/Contrasting-Patterns-in-Crop-Domesticationand?searchresult=1#usercomments

Winters C (2010). The Kushite Spread of haplogroup R1*-M173 from Africa to Eurasia. Current Research Journal of Biological Science 2(5) 294-299. Available: http://maxwellsci.com/print/crjbs/v2-294-299.pdf

Winters C (2010b). A Sub-Saharan Origin of the Early European Farmers. Comment: Ancient DNA from European Early Neolithic Farmers Reveals Their Near Eastern Affinities, by Wolfgang Haak et al., PLOS Biology, November 9, 2010, https://doi.org/10.1371/journal.pbio.1000536

Winters C (2010c). The Fulani are not from the Middle East. Proceedings of the National Academy of Sciences of the United States of America. 107(34):E132. doi:10.1073/pnas.1008007107

Winters C (2010d). Sickle Cell Anemia in Africa and India.

http://www.ispub.com/journal/the_internet_journal_of_h
ematology/volume_7_number_1_40/article/sickle-   cell-
anemia-in-india-and-africa.html

Winters C (2010e). Y-Chromosome evidence of African Origin of Dravidian Agriculture. http://www.academicjournals.org/ijgmb/PDF/pdf2010/Mar/Winters.pdf

Winters C (2011). Possible African origin of Y-Chromosome R1-M173. International Journal of Science and Nature 2(4) 743-745. Available: http://www.scienceandnature.org/IJSN_Vol2(4)D2011/IJSNVOL2(4)-9.pdf

Winters C (2011b). First European Farmers were not Eastern Europeans. Webmed Central Human Genetics 2(9). Available: http://www.webmedcentral.com/article_view/2265

Winters C. (2012). Origin of the Niger-Congo Speakers. Webmed Central Genetics, 3(3) WMC003149doi: 10.9754/journal.wmc. 003149

Winters C (2014). Ancient History of the Tamils in Central Asia. Createspace.

Winters C. (2014b). Reader Comment, Ethio-Semitic People Took "Eurasian" Genes to South Arabia and the Levant: Hodgson JA, Mulligan CJ, Al-Meeri A, Raaum RL, Early Back-to-Africa Migration into the Horn of Africa. PLoS Genet 10(6): e1004393. https://doi.org/10.1371/journal.pgen.1004393

Winters C (2016). The Phylogeography of Afro-Americans and Africans, (Createspace, Amazon).

Winters C (2017). Did African Slaves Bring the Y-Chromosomes R1 Clades to the Americas? International Journal of Innovative Research and Review 5(2) 1-10. Available: http://www.cibtech.org/JInnovative-Research-Review/Publications/2017/VOL-5-NO-2/01-JIRR-001-

JUNE-WINTERS-DIDCHROMOSOME.pdf

Winters C (2017b). A GENETIC CHRONOLOGY OF AFRICAN Y-CHROMOSOMES R-V88 AND R-M269 IN AFRICA AND EURASIA . Indian Journal of Fundamental and Applied Life Sciences, 7(2) 24-37.
Winters C (2017c). A King's Seal? Was Pharaoh Apophis Originally King of the Mythical Kushites? http://www.ancient-origins.net/history-famous-people/king-s-seal-was-pharaoh-apophis-originally-kingmythical-kushites-008430?nopaging=1
Winters,C. (2018). The Kushites, Who,What, When Where.
Wood ET, Stover DA, Ehret C, Destro-Bisol G, Spedini G, McLeod H, Louie L, Bamshad M,Strassmann BI, Soodyall H and Hammer MF (2005). Contrasting patterns of Y-chromosome and mtDNA variation in Africa: evidence for sex-biased demographic processes. European Journal of Human +Genetics 13 867-876.

Caucasian and African Relations before The Atlantic Slave Trade

The foundation of population genetics is that differences in genetic ancestry correlate into the "social construct" of real races (Reich, 2018). These biological differences correspond to at least three or four continental human populations: Sub-Saharan African (Negro), Eurasian (West (Europeans) and East Asia (Mongoloid) and Native American. Thusly, these continental populations only carried a select number of haplogroups.

Dr. Reich (2018) of Harvard University wrote that "Groundbreaking advances in DNA sequencing technology have been made over the last two decades. These advances enable us to measure with exquisite accuracy what fraction of an individual's genetic ancestry traces back to, say, West Africa 500 years ago—before mixing in the Americas of the West African and European gene pools that were almost completely isolated for the last 70,000 years". This view of human history that Africans and Europeans only began mixing 500 years ago by population geneticists promotes the idea that human populations have always inhabited the continental regions where you find them today since the first out of Africa event 60,000 years ago. And, that Eurasian only came in contact with Sub-Saharan Africans (SSA) during the Atlantic Slave Trade.

This view of human biological history is false. Historical and archaeological research indicates that Eurasians and SSAs were not isolated from each other; and have been in physical contact for thousands of years.

This was especially true of Caucasians after they began to migrate into Europe from Central Asia 3500 years ago. The Caucasians who are the contemporary Europeans, have had considerable interaction with Negroes or SSAs that have been living in Europe for past 40,000 years. This has resulted in the mixing of the African and Eurasian races.

Thusly, human races cannot be separated into neatly aggregated human populations carrying a unique package of genes specific to the continents they now inhabit because these continents were originally inhabited by Negroes or SSAs, when Caucasians and Mongoloid people came in contact with Black or Negro people 3500 years ago in Eurasia, and 500 years ago in the Americas.

The oldest skulls found in Spain are of Africans. In ancient and pre-Medieval times southern Spain and Portugal were Africoid lands.

Magyar

The Magyar or Hungarian people trace their origin back to Nubia. According to Dr. Vamos Toth Bator, in ages past the Magyar were members of a worldwide civilization he calls Tamana. This is proven by the thousands of common toponyms shared by people around the world on different continents.

The Magyar came directly from the Proto-Sahara in ancient times. They early settled Central Asia until the Chinese drove the Kushana out of Gansu Province. The ancient Magyar had their own writing system which was based on the Proto-Saharan script. The runes used by the Magyar were also based on this writing.

Other early Magyar settled in the Carpathian mountains. They are remnants of the early Negro populations that lived in this area 4000 years ago. The term *Ugrian*, for the early Hungarians indicates that they were dark skinned. The Hungarian language is genetically related to Dravidian and Black African languages.

Celts

The most ancient culture on the European Peninsula was that of the *Afro=Iberians* or *Celtiberians*. These "Black Celts " were described by Tacitus , a Roman as *"swarthy with curly hair"*. The Basques may be descendants of this African population. The Basque language is closely related to Dravidian and African languages.

The Celts were originally Black people. Ephorus (c. 405BC) claimed that the Celts were Blacks or Ethiopians(1)The Celts continued to be recognized as Blacks by Tacitus, who wrote about the Black Celts and Picts in 80 AD[1].(2)

The Celts on the mainland of Europe were called Iberians or Silures[2].(3)though the original Celts were Black, overtime their name was stolen by Europeans. Father

---

[1] J.A. Rogers, *Sex and race*, Vol.1 (New York, 1967) p.196.

[2] *Ibid,* p.196.

O'Growney has discussed the history of the Celts. He makes it clear that the original Celts were the Iberians[3].(4)

The Iberians were probably conquered by the Ligurians[4].(5) Some historians have suggested that the Ligurians may be represented by the modern Basque of Spain[5].(6) The Ligurians took the name Celt.

The Ligurians/Celts were conquered by the Gaulish speaking people. The Gauls conqered the Ligurians and pushed them into Spain. It was these Gauls who imposed their language on the Iberian and Ligurian Celts .

The Gauls were Belgians according to Father O'Growney. The Irish and Welsh are descendants of these Gauls[6].(7)The Gauls spoke Gaulish or Gaelic.(8) The Germans conquered the Gaulish-Celts, and Gaulish disappeared around 4th Century AD.

All of the Black Celts in Britain were not erased by the Gauls. This is supported by the Ivory Lady of York ,England. The reconstruction of the face of the Ivory Bangle Lady (c.350AD) indicates that she was African or Black. This woman was rich and indicates the African type common to the Bristish Isles.

There is genetic and linguistic evidence that proves that the Celts were Black or African people. An examination of the language spoken by the Basque has a Niger-Congo

---

[3] Father O'Growney, The Celts, *American Ecclesiastical Review*, (1901) pp.350359.
[4] *Ibid*, p.352.
[5] *Ibid.,* p.353.
[6] *Ibid.*, p.352.

substratum. C.J.K. Cambell-Dunn has found a Niger-Congo substratum in Basque[7].(9) Dr. Cambell-Dunn found that the Niger-Congo and Basque languages share personal pronouns, numerals and vocabulary items.

There is also genetic evidence linking the Basque and Niger-Congo speakers. Both groups share SRY10831.1, YAP, M2,M173(xR1a,R1b3), E3*-P2, E3b2-M81[8]. (10) This linguistic and genetic evidence supports the African origin of the Celts.

Black Greeks

The Minoans, who were Africans introduced Linear A, whose signs are identical to the writing left by Africans throughout the Sahara, like those found at Tichitt and presently represented in the Vai and several other West African scripts.

The Afro-Greeks were not called Ethiopian. They were native to Greece when the Indo-Europeans came to Greece. Homer, who was an Afro-Greek referred to his people as *xanthos* (brown) in color.

The earliest inhabitants of Greece and the Aegean Islands were Blacks from ancient Libya, Palestine, and Asia Minor. These Blacks founded Athens, Thebes Thera and Attica. They occupied much of the mainland and all the Aegean

---

[7] C.J.K. Cambell-Dunn, *Basque as Niger Congo*. Retrieved 8/28/2010 at http://home.clear.net.nz/page/gc_dunn/Basque_as_Niger-Congo.html

[8] Alonso S, Flores C, Cabera V. The place of Basque in the European y-Chromosome diversity landscape, *European Journal of Human Genetics,* (2005) 13:pp.1293-1302.

Islands.

This is one of the Thera Frescos. Note the busy atmosphere Associated with the Pelasgian cities during the 16th Century BC.

These Blacks are frequently depicted in the art associated with the so-called Dark Ages (1200-600 BC). There are also fine frescos from Thera (Sanorin) Island which illustrate one of the Aegean cities occupied by these Blacks during the 16th and 15th centuries BC.

Although these people of the Heroic age came from diverse origins, the Aryan-Greeks called them Pelasgians. According to the Greeks, the first man was Pelasgus-- ancestor of the Pelasgians. The Pelasgians were a combination of different Black tribes called *Achaeans, Cadmeans, Leleges, Carians* or *Garamantes*.

The term Pelasgian was applied to all these pre-Hellenic inhabitants of Greece. R.J. Hopper, in **The Early Greeks**, noted that "indeed the classical Greeks believed in the separate existence of diverse ethnic elements side by side, and thought particularly of the Pelasgians in this connection".

According to tradition, the *Pelasgians* inhabited Arcadia and many Aegean Islands. These Blacks took their own writing to Greece which was later used by the Aryan-Greeks. According to Herodotus quadrigas or four-horse chariots were introduced to Greeks by the Libyans .

The caucasian-Greeks adopted the language of the Pelasgians and Egyptians. The linguistic evidence shows that there was a differentiation of Greece into East Greek and West Greek. The Black Greeks spoke East Greek (*Achaioi* or *Achaean*). West Greek was spoken by the Dorian or Aryan Greeks. The earliest Aryan tribe called Ionians spoke a dialect of East Greek called Aeolic.

Many classical scholars teach the world that the Greek language is entirely Indo-European. This view of Greeks is wrong.

Dr. Anna Morpurgo Davies, has made it clear that "less than 40% of the words which have an Indo-European etymology". According to Dr. Davies, 52.2 % of the Greek terms in *Chantraine's Dictionnaire Etymologique de la langue Grecque* (1968) have an unknown etymology. The mixed nature of the Greek language results from the early settlement of the Aegean by Blacks from Africa.

Some of these words are of African origin. Robert K.G. Temple, in **The Sirius Mystery**, shows that many of the most common words of the Greek vocabulary are of Egyptian origin. Diop (1991) has also discussed the

Egyptian origin for many Greek terms.

The Xanthos or Palasgians of Thera

The Greeks often called the first inhabitants of Greece Pelasgians. The Greek writers claimed that Pelasgus, the great ancestor of the Pelasgians was the first man. The Pelasgians were a combination of diverse Black tribes which included the Achaeans , Kadmeans, and Leleges. The Garamantes were also often called Pelasgians by some classical writers. Strabo said "that the Pelasgi, as indeed the most ancient nation, were diffused through all Greece, and especially among the Aeolians".

The city of Argo was founded by Phoroneus, the father of Pelasgus, Iasus and Agenor. It was these folks who divided the Peloponnese between them.

Herodotus referred to the *Pelasgians* as "venerable ancestors". He said that the first Athenians "they were Pelasgi, the later possessing the country now designed Hellas". The Pelasgian founding of Athens is also noted by Plutarch in Theseus 12, and Ovid in **Metamorphosis** vii.402ff.

According to Herodotus vii.91, the Pelasgians also founded Thebes in Europe. Pausanias, noted that "The Arcadians make mention of Pelasgus as the first person who existed in their country. From this king the whole region took the name Pilasgia". Hopper noted that the Pelasgians founded Attica.

The Black immigrants from Canaan were also settled in the Aegean at Argolis. They called themselves the *"Sons of Abas"*. Many of the *Melampodes* later took part of Argolis away from the Canaanites.

The earliest Greek alphabet was made by the Pelasgians, it was lost and later reintroduced by Kadmus to Boeotia. Another Pelasgian, Evander of Arcadia introduced writing to the Italians. This script was used to make the first fifteen characters of the Latin script according to Pliny and Plutarch.

Pelasgians from Thera

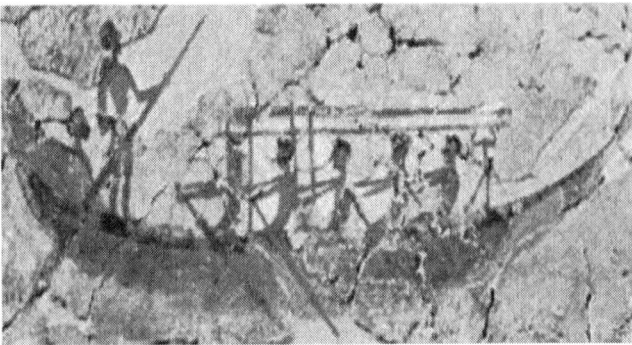

Pliny says that one of the Aegean scripts was created by an Egyptian named Menos. An Egyptian creation of one of the early Greek alphabets is not out of the question because the early Predynastic Egyptians used the Proto-Saharan script as did the founders of the 12th Dynasty. Moreover, the Tiles of Rameses II, published by F. Hitching, in **The Mysterious World**, are analogous to the early Greek characters. Europeans adopted this writing to write business documents and we know it as Linear B. Europeans only got writing from the Egyptians. The Caucasian Greeks who obtained writing from the Blacks of Africa and Phoenicia passed on writing to the Romans. With the fall of Rome Western Europeans got writing from the African Muslims who taught them the arts and sciences.

The original Danes or Vikings were Blacks[9]. (11)This is made clear in the Oseberg 8th Century Vikings on the Norway Sledge carving of the Black seafarers that populated the region at this time. It is clear from this carving that the 8th Century Vikings were different from the Blond, big bodied folk of Viking legends.

In 711 African Muslims from Senegal and the Western Sudan under the leadership of Tarik established the first great civilization in Europe. The second wave of Blacks to enter Spain came from an area around the Upper Senegal River. These Africans and Arabs are known in history as the *Almoravids*.

Their leader Yusuf, according to the "***Roudh el Kartar***, was a pure African. In Spain the Almoravids built Universities and taught European Christians the arts and sciences of African and Greek origin. These Africans ruled at Grenada until 1492.

---

[9] W.B., The Doctrine of Celtism, *Notes and Queries*, (1871) 7: p.8.

German Barbarians Attacking a Strasburg Castle

The African Moors ruled much of Europe from Iberia up to modern-day France and Germany. They built the first centers of learning in Europe and re-introduced sciences to the Caucasians. These Moors, as they were called also occupied large parts of Italy. In 846 the African Muslims seized Rome itself.

This short review of European history makes it clear that Reich (2018) was wrong when he said Africans and Europeans had been living in isolation for 70,000 years. History indicates Europeans and SSA have been in intimate contact for 1000's of years. As a result the idea that Europeans and Africans only came into contact 500 years ago is groundless.

References:
1. William and Robert Chambers, *Chambers Information for the people*, Vol. 2 ,(London & Ediburgh,1884) p.66.

2 J.A. Rogers, *Sex and race*, Vol.1 (New York, 1967) p.196.

3. *Ibid,* p.196.

4. Father O'Growney, The Celts, *American Ecclesiastical Review*, (1901) pp.350359.

5. *Ibid*, p.352.

6. *Ibid.,* p.353.

7. *Ibid.*, p.352.

8. *Ibid.*, p.353.

9. C.J.K. Cambell-Dunn, *Basque as Niger Congo*. Retrieved 8/28/2010   at http://home.clear.net.nz/page/gc_dunn/Basque_as_Niger -Congo.html

10. Alonso S, Flores C, Cabera V. The place of Basque in the European y-Chromosome diversity landscape, *European Journal of Human Genetics,* (2005) 13:pp.1293-1302.

11. W.B., The Doctrine of Celtism,*Notes and Queries*, (1871) 7: p.8.

12. David Reich, How Genetics is Changing Our Understanding of 'Race'. New York Times, 23 March 2018.

First Americans were Africans

The Americans was also early settled by Negroes or Sub-Saharan Africans. As a result, there is no way geneticists can claim Monglod Native Americans and Africans only came in contact 500 years ago. Archaeological evidence indicates that anatomically modern humans (amh) Negroes or Blacks were in South America tens of thousands of years before Ice Age people could have crossed a viable land bridge between Alaska and Siberia 13,000 years ago.

There is no way anyone can claim that there were no Black Native Americans before 1492, because even the Spanish said the first Indians they met were Black people like the Africans and people of South Indian. Neither genetic evidence nor craniometrics deny the existence of Black Native Americans.

The Native Americans were called Indians because they were Black skinned like the Natives of South India. As I have noted above Quatrafages noted the numerous American tribes that were Negro Native Americans.

Craniometric quantitative analysis and multivariate methods have determined the Native American populations. This research indicated that the ancient

Americans represent two populations, paleoamericans who were phenotypically African, Australian or Melanesian and **a mongoloid population that appears to have arrived in the Americas after 6000 BC.**

The determination of the Paleoamericans as members of the Black Variety is not a new phenomena. Howells (1973, 1989, 1995) using multivariate analyses, determined that the Easter Island population was characterized as Australo-Melanesian, while other skeletons from South America were found to be related to Africans and Australians (Coon, 1962; Dixon, 2001; Howell, 1989, 1995; Lahr, 1996). The African-Australo-Melanesian morphology was widespread in North and South America. For example skeletal remains belonging to the Black Variety have been found in Brazil (Neves, Powell, Prous and Ozolins, 1998; Neves et al., 1998), Columbian Highlands (Neves et al., 1995; Powell, 2005), Mexico (Gonza'lez-Jose, 2012), Florida (Howells, 1995), and Southern Patazonia (Neves et al., 1999a, 1999b).

We don't have to depend on just paintings to acknowledge the Negro/African presence in America before 1492, we also have the facial reconstructions of paleoAmericans that have resulted from craniometrics that show these people were Blacks. The bioanthropologist Walter Neves's reconstruction of the first Americans evidenced Negroid features for the Paleoamerican we call Luzia. What made this finding startling was that Neves using the mahalanobis distance and principal component analysis, found that 75 other skulls from Lagos Santa, were also phenotypically African or Australian (Neves et al., 2004).

Researchers should stop trying to claim there were no

Blacks in America before 1492, Blacks had been in America 94,000 years according to Dr. Nieda Guidon before the mongloid Native Americans found in America today arrived in the United States 6000 years ago.

Given the fact that the earliest dates for habitation of the American continent occur below Canada in South America is highly suggestive of the fact that the earliest settlers on the American continents came from Africa before the Ice melted at the Bering Strait and moved northward as the ice melted.

 An African origin for these people is a good fit because Ocean Currents would have carried migrants from Africa to the Americas, since there was no Ice Age sheets of ice to block passage across the southern Atlantic.

Dr. Bryan (1987) noted many sites where PaleoAmericans have left us evidence of human habitation, including the pebble tools at Monte Verde in Chile (c.32,000 B.P), and rock paintings at Pedra Furada in Brazil (c.22,000 B.P.) and mastodont hunting in Venezuela and Colombia (c.13,000 B.P.). These discoveries have led some researchers to believe that the Americas was first settled from South America.

    The main evidence of the ancient Americans are prehistoric tools and rock art, like those found by Dr. Nieda Guidon (1986.1991, 1996).

Today archaeologists have found sites of human occupation from Canada to Chile that range between 20,000 and 100,000 years old (NYT,2015). Dr.Nieda Guidon (1986.1991, 1996), in numerous articles claims that

Africans were in Brazil between 65,000-100,000 years ago. Dr. Guidon claims that man was at the Brazilian sites for early Americans, 65,000 years ago. She told the New York Times, that her dating of human populations in Brazil 100,000 years ago , was based on the presence of ancient fire and tools of human craftsmanship at human habitation sites.

P.S. Martin and R. G. Klein (1989) after discussing the evidence of mastodont hunting in Venezuela 13,000 years ago observed that : "The thought that the fossil record of South America is much richer in evidence of early archaeological associations than many believed is indeed provocative.... Have the earliest hunters been overlooked in North America? "

Warwick Bray (1987), has noted that there are numerous sites in North and South America which are over 35,000 years old. A.L. Bryan noted that these sites include , the Old Crow Basin (c.38,000 B.C.) in Canada; Orogrande Cave (c.36,000 B.C.) in the United States; and Pedra Furada (c.45,000 B.C.) .

Using craniometric quantitative analysis and multivariate methods, Dr. Neves (1989,1990, 1991, 1999b), determined that Paleo Americans were either Australian, African or Melenesians . The research of Neves indicated that the ancient Americans represent two populations, PaleoAmericans who were phenotypically African, Australian or Melanesian and a mongoloid population that appears to have arrived in the Americas after 6000 BC.

Archaeologist have reconstructed the faces of ancient Americans from Brazil and Mexico. These faces are based on the skeletal remains dating back to 12,000BC. The PaleoAmericans resemble the first Europeans.

Naia   First European   Luzia

PaleoAmericans and First European

Researchers working on the prehistoric cultures of these ancient people note that they resemble the Black Variety of humanity, instead of contemporary Native Americans. The Black Variety include the Blacks of Africa, Australia, and the South Pacific.

Dr. Chatters who found Naia's skeleton, told Smithsonian Magazine that: "The small number of early American specimens discovered so far have smaller and shorter faces and longer and narrower skulls than later Native Americans , more closely resembling the modern people of Africa, Australia, and the South Pacific. "This has led to speculation that perhaps the first Americans and Native Americans came from different homelands," Chatters continues, "or migrated from Asia at different stages in their evolution."

Although Dr. Chatters believes the PaleoAmericans came from Asia this seems unlikely, because of the Ice sheet that blocked migration from Asia into the Americas. C. Vance Haynes noted that: "If people have been in South America for over 30,000 years, or even 20,000 years, why are there so few sites?....One possible answer is that they were so few in number; another is that South America was somehow initially populated from directions other than north until Clovis appeared".

The fact that the Beringa land bridge was unviable 15kya make it unlikely that during the Ice Age man would have been able to walk or to sail from Asia to South America at this time. As a result, these people were probably from Africa as suggested by Dr. Guidon.

In summary, the land bridge between Siberia and Alaska was unviable before 13,000 BC. Eventhough man could not enter the Americas until after 14kya, man was probably in South America as early 100kya according to Dr. Guidon's research in Brazil.

The first people in the Americas are called PaleoAmericans. The research of Chatters and Neves indicate that the PaleoAmericans were not mongoloid. These researchers claim the PaleoAmericans " more closely resembl[ed] the modern people of Africa, Australia, and the South Pacific."

The first Americans probably came to the Americas by sea, due to the unviable land route to the Americas before 13,000 BC. As a result, we must agree with Dr. Guidon that man probably traveled from Africa to settle prehistoric America.

The Archaeological evidence indicates that PaleoAmericans settled South America before North America, and that these Americans did not belong to the Clovis culture. Africa is the most likely origin of the PaleoAmericans, because the Ice sheet along the Pacific shoreline of North America, Siberia and Alaska, would have made the sea route from Asia or Europe unviable 65,000 years ago. The Dafuna boat dating back to 8,000 BC, shows that Africans had boats at this early date. The culture associated with the Dafuna boat dates back to 20kya.

References:

Bray, Warwick. 1988. "The Paleoindian debate". <u>Nature</u> 332, (10 March), p.107.
Bryan, A. L. 1987. "Points of Order". <u>Natural History</u> , pp.7-11.
Coon CS (1962). The Origin of Races (New York: Knopf).

Dixon EJ (2001). Human colonization of the Americas: timing, chronology and process. Quaternary Science Review 20 277–99.

Gonza´lez-Jose´ R, Hernande´z M, Neves WA, Pucciarelli HM and Correal G (2002). Cra´neos del Pleistoceno tardio-Holoceno tempramo de Me´xico en relacio´n al patro´n morfolo´gico paleoamericano. Paper presented at the 7th Congress of the Latin American Association of Biological Anthropology, Mexico City.

Guidon, N. and Delibrias, G. 1986. "Carbon-14 dates point to man in the Americas 32,000 years ago." Nature 321:769-

771.

Guidon, N., and B. Arnaud. 1991. "The chronology of the New World: Two faces of one reality." World Arch. 23(2):167-178.

Guidon, N., et al.1996. "Nature and Age of the Deposits in Pedra Furada, Brazil: Reply to Meltzer, Adovasio & Dillehay," Antiquity, 70:408.

Haynes,Jr., C.V. 1988. "Geofacts and Fanny". Natural History ,(February)pp.4-12.

Howells WW (1973). Cranial Variation in Man: A Study by Multivariate Analysis of Patterns of Difference among Recent Human Populations, Papers of the Peabody Museum of Archaeology and Ethnology (Cambridge, MA: Harvard University) 67.

Howells WW (1989). Skull Shapes and the Map: Craniometric Analyses in the Dispersion of Modern Homo, Papers of the Peabody Museum of Archaeology and Ethnology (Cambridge, MA: Harvard University) 79. Early Holocene human skeletal remains from Cerca Grande 497

Howells WW (1995). Who's Who in Skulls: Ethnic Identification of Crania from Measurments, Papers of the Peabody Museum of Archaeology and Ethnology (Cambridge. MA: Harvard University) 82.
Kumar, Mohi. 2014. DNA From 12,000-Year-Old Skeleton Helps Answer the Question: Who Were the First Americans? Retrieved 16 August 2016 at :
http://www.smithsonianmag.com/science-nature/dna-12000-year-old-skeleton-helps-answer-question-who-were-

first-americans-180951469/?no-ist

Martin, P. S. and R.G.Klein (eds.),Quarternary Extinctions: A Prehistoric Revolution, (Tucson:University of Arizona Press,1989) p.111.

Neves, W. A. and Pucciarelli, H. M. 1989. Extra-continental biological relationships of early South American human remains: a multivariate analysis. Cieˆncia e Cultura, 41: 566–75

Neves, W. A. and Pucciarelli, H. M. 1990. The origins of the first Americans: an analysis based onthe cranial morphology of early South American human remains. American Journal of Physical Anthropology, 81: 247.

Neves, W. A. and Pucciarelli, H. M. 1991. Morphological affinities of the first Americans: an exploratory analysis based on early South American human remains. Journal of Human Evolution, 21: 261–73.

Neves, W. A. and Meyer, D. 1993. The contribution of the morphology of early South and Northamerican skeletal remains to the understanding of the peopling of the Americas. American Journal of Physical Anthropology, 16 (Suppl): 150–1.

 Neves, W. A., Powell, J. F., Prous, A. and Ozolins, E. G. 1998. Lapa Vermelha IV Hominid 1: morphologial affinities or the earliest known American. American Journal of Physical Anthropology, 26(Suppl): 169.

 Neves, W. A., Powell, J. F. and Ozolins, E. G. 1999a. Extra-continental morphological affinities of Palli Aike, southern Chile. Interciênncia, 24: 258–63.

Neves, W. A., Powell, J. F. and Ozolins, E. G. 1999b. Modern human origins as seen from the peripheries. Journal of Human Evolution, 37: 129–33.

Neves W.A . and Pucciarelli H.M. 1991. "Morphological Affinities of the First Americans: an exploratory analysis based on early South American human remains". Journal of Human Evolution 21:261-273.

Neves W.A ., Powell J.F. and Ozolins E.G. 1999. "Extra-continental morphological affinities of Lapa Vermelha IV Hominid 1: A multivariate analysis with progressive numbers of variables. Homo 50:263-268

Neves W.A ., Powell J.F. and Ozolins E.G. 1999. "Extra-continental morphological affinities of Palli-Aike, Southern Chile". Interciencia 24:258-263.
http://www.interciencia.org/v24_04/neves.pdf

Neves, W.A., Gonzá lez-Jose´ , R., Hubbe, M., Kipnis, R., Araujo, A.G.M., Blasi, O., 2004. Early Holocene Human Skeletal Remains form Cerca Grande, Lagoa Santa, Central Brazil, and the origins of the first Americans. World Archaeology 36, 479-501

Neves, W. A., and M. Hubbe. 2005. Cranial morphology of early Americans from Lagoa Santa, Brazil: Implications for the settlement of the New World. Proc. Natl. Acad. Sci. USA 102:18,309–18,314.

NYT (New York Times). (2015) Human's First Appearance in the Americas .

http://www.nytimes.com/2014/03/28/world/americas/discoveries-challenge-beliefs-on-humans-arrival-in-the-americas.html?hp&_r=4

Powell,J.F. (2005). First Americans:Races, Evolution and the Origin of Native Americans. Cambridge University Press.

Winters,C. (2013). African Empires in Ancient America. https://www.amazon.com/African-Empires-Ancient-America-Winters/dp/0615796583
Winters,C. (2015). THE PALEOAMERICANS CAME FROM AFRICA,*jirr. Vol. 3 (3) July-September, pp.71-83/Winter.*
https://www.academia.edu/17137182/THE_PALEOAMERICANS_CAME_FROM_AFRICA

Manufactured Genetic Origins reveal Geneticists have created a fake origin for the Indo-Europeans

The research findings in this book reveal that the agro-pastoral people who founded civilization in Europe were Kushites from the Nile Valley. The proceeding chapters of **Manufactured Genetic Origins** have made it clear that 1) the Anatolian and Steppe agro-pastoralist were not Indo-Europeans; and 2) Eurasians and Africans have been mixing and in contact for thousands of years up to , and after the Atlantic Slave Trade.

The foundation of population geneticists that the races were isolated for thousands of years was disconfirmed by this herein. Population geneticists use genomic evidence and Bayesian statistics to infer prehistoric demographic events without archaeology is a fruitless intellectual effort. Colin Refrew (2000) made the point two decades ago when he advocated using archaeology, linguistics and paleo-anthropological (ALPA) evidence as a method to reconcile the findings of population genetics with corroborated information from the ALPA fields. Some geneticists attempted to use ALPA evidence to support their findings.

Other geneticists failed to use ALPA evidence to corroborate their findings and statistical-significance test to confirm or disconfirm their findings. And by 2010, many geneticists decided that Bayesian statistics alone could support their research even if ancient DNA (aDNA) was unavailable.

The failure of geneticists to make hypotheses and use statistical significance tests, to confirm or disconfirm hypotheses—instead of Bayesian statistics—makes their inferences about population movements and identification of "races" mere speculation. This results from the fact that Bayesian statistics reflect the opinion about a data set, already held by the researcher. Consequently when geneticists makes their research questions, the focus of their research ,without hypotheses, will always make the resulting answer to the research question true—even when the authentic answer should be false.

The use of Bayesian statistics has made it possible for geneticists to use spurious spatial autocorrections to enforce trends in the data that support their research questions. Thusly, Bayesian equations, begin with only one prior variable, and offer the researcher confirmation of his research question—but no real evidence that the research question was a valid and reliable predictor of prehistorical events because a statistical model of a historical event remains hypothetical.

The genome data in population genetics research articles is valid and reliable evidence of the people, or bones from which the sample was taken. But this is descriptive data of the genome within and among the population under study, but this genomic data cannot inform historical events except as evidence of the genome carried by the people used in the sample.

Reich (2018b) wrote that "the ancestors of East Asian, Europeans, West Africans and Australians were until recently, almost completely isolated from one another for 40,000 years or longer, which is more than sufficient time for the forces of evolution to work". The archaeology and history outlined in **Manufactured Genetic Origins** , of Eurasian prehistory proves that this statement by Reich (2018,2018b) is false.

The interaction between Eurasian and Africans forever, makes it impossible for geneticists to claim that there were three distinct continental populations. Absent corroborating evidence that the populations were "isolated" until recently makes the claim by some geneticists that although race is a "social construct", genomics can identify varied races is false.

It is false because Blacks or SSAs have always inhabited Eurasia and the Americas, so there is no such thing as continental specific genomes or haplogroups. The reality that SSAs have always been in Eurasia, the dating of specific Eurasian haplogroups is impossible, because there is no archaeological or craniometric evidence of Caucasians or mongoloids in Eurasia prior to 3500 kya.

Eventhough, David Reich, in **Who we are and How we Got Here** , argues that "the geneome revolution is the study of the human past" which allows population geneticists to write history using ancient DNA (aDNA), the archaeological evidence discussed herein, makes it clear he was wrong. Reich (2018) was wrong because the varied populations have been in constant communication—not isolated from one another.

The research into Indo-European origins by geneticists indicate that they believe in a fallacy. The fallacy is that geneticists believe they can write history solely on Bayesian statistics.

Lewis-Kraus (2019) says Reich (2018) argues that the aDNA Revolution "allows geneticists with greater detail and certainty to study human history, evolution and identity than previous historical methods like archaeology or history. These researchers believe that archaic bones supplies the genome that after analysis via "expensive sequences" : and Bayesian statistical techniques provide us with thousands of human prehistory.

But these statistical methods have not superceed archaeology as I have shown in the previous chapters of **Manufactured Genetic Origins** . The findings in this book prove that the Steppe and Anatolian agro-pastoralists were Kushites—not Indo-Europeans.

This finding was not unexpected. The manufactured history of the Indo-Europeans by Reich et al, was going to end in failure once the geneticists abandoned the archaeogenetic method for total reliance on Bayesian statistics to model prehistory.

The potential for failure in using Bayesian statistics to write prehistory did not go unrecognized. Lewis-Kraus (2019) noted that geneticists are harvesting aDNA and making revisions of prehistory which may not be accurate. Much of the controversy comes from the reality that the major contemporary population geneticists ignore archaeological evidence and depends statistical analysis of genomic data to make their inferences concerning prehistory.

The failure of geneticists to use archaeological data made their guesses about Indo-European prehistory wrong.They wrong because geneticists had no way to corroborate their findings with hard physical evidence.

Archaeologists rely on three sources when they cross-reference prehistorical events. These sources are 1) ancient written records, 2) sedimentary chronology and 3) tree and radio-carbon dating (Lewis-Kraus, 2019). The absence of cross referencing standards for Bayesian statistical models, is what led to geneticists creating a manufactured , fake history of the Indo-Europeans.

References:
Lewis-Kraus,G. (2019). Is ancient DNA Research Revealing New Truths—or Falling into Old Traps? The New York Times ,17 January 2019.

Reich, D. (2018). Who We Are and How we Got Here.

Reich, D. (2018b). How Genetics Is Changing Our Understanding of "Race". New York Times 23 March 2019, downloaded 12/4/19

A Protocol to Evaluate Population Genetics Papers

The most important activity of the researcher is reading research articles. Although many population genetics articles are published graduate students and professional geneticists need a tool to evaluate the articles to determine their relevance. Here we provide an evaluation tool researchers can use to evaluate genetics research articles. There are tens of articles published each year in population genetics. These articles are must reading for anthropologist and molecular geneticists interested in migration and population genetics.

Using Bayesian statistics molecular geneticists make inferences about prehistoric demographic events, relating to various ethnic populations. To reconcile their genomic evidence with prehistoric and historical information some population geneticists use archaeological, linguistic and paleoanthropological data to corroborate their DNA findings.. The use of use archaeological, linguistic and paleoanthropological data to support molecular genetics is called archaeogenetics (Renfrew and Boyle,2000).

Geneticists and anthropologists use Archaeogenetics to explain and discuss past population events. Archaeogenetics can be defined as the use of prehistoric and historical events to determined by archaeology, genetics and linguistics in concert with the DNA of various ethnic groups to infer the ethnic identity of ancient populations and/or the ancient migration of one population to another geographical location.

Method

The research design used in this paper is a literature based research methodology. We sampled the archaeogenetics literature base. The literature was analyzed to determine what criterions explain best the relationship between contemporary DNA, ancient DNA human remains and past population events within the context of the "wave of advance" model. The top criterions used to evaluate research papers were used to construct this Checklist.

Results

In population genetics the researcher usually uses the "wave of advance" model to explain demographic movements in the past. The "wave of advance" model was used to explain the spread of advantageous genes within a population(Ackland et al,2007; Renfrew, 2001).) . This theory was adapted to explain why an advantageous technology that may appear in one population spreads (and or taken)to another population living in a different geographical area (Ackland et al, 2007).

Although archaeogenetics is the norm for many molecular geneticists, most researchers believe that Bayesian statistics alone, have sufficient power to demonstrate the valility of their research, and fail to corrobate the DNA data with corresponding archaeological, linguistic and paleo-anthropological evidence.

Many people don't know how to evaluate population genetics articles, because they are expost facto research based on " statistical infererences" or the beliefs of the researcher supported by statistics. As a result, researchers can not judge the validity and reliability of the research. One must assume the research is correct based solely on the Bayesian statistical inferences—not the interactions between an independent variable and dependent variable(s). In research there are two variables: variables that can be manipulated and variables that can not be manipulated.

A variable that can be manipulated is a variable that can be changed for example, your ability to perform a particular task can be influenced by the amount of training you receive in performing the task.

A variable that can not be manipulated can not be changed . For example, right now you are a particular age, it can not be manipulated. You are either Black or white, race can not change.

Research studies include a number of variables. Variables which can be manipulated or not manipulated.

Independent Variable (IV) any variable used to control for individual differences (this variable usually not manipulated). The Dependent variable (DV) any outcome measure which is effected by the IV. For example, the effect of sex (IV) on reading achievement (DV).

Validity is testing the appropriateness, meaningfulness and usefulness of specific inferences made from test scores. In qualitative research the extent to which the research uses methods and procedures that ensure a high degree of research quality and rigor.

Internal Validity, we assume that whatever was manipulated produced a change in the dependent measure. The IV insures by control of the extraneous variables: health, sex, race, SES, age, IQ, religion.

External Validity, provides the ability to generalize the findings. In other words the IV produced a change in DV.
In normal scientific research the researcher states a hypothesis and uses the scientific method to test his/her hypothesis. The validity and reliability of the piece of research is then determined by statistical significance tests focused on the interaction between the independent and dependent variable.

In the traditional evaluation of a piece of research literature you look at the researcher's hypothesis, results and statistical methods s/he used to determine the statistical significance of the research. This is not the case in population genetics research; in this research you are evaluating statistical inferences based on *the beliefs already held by the researcher* about a set of data, instead of testing a hypothesis.

As a result, the research contained in a population genetics article, reflects the views and beliefs already held by the researcher. Thusly, the statistical inferences will automatically support the views and beliefs held by that researcher; and any outliners that fail to support the researcher's beliefs may not be mentioned in the resulting research article/paper.

Here we will ask the question: "How do you evaluate population genetics research if it is expost facto research, that lacks an experimental design?" First, we will attempt to look at the doxa that may influence a geneticist's research and the constructs that should be considered when evaluating this knowledge base.

In reading any piece of research literature, we assume that any article or book written by an establishment member of the academe is reliable and valid. A piece of research full of valid scientific and/or historical truths-- erudite scholarship and impeccable research based on the scientific method.

The scientific method is based on hypotheses testing. Hypotheses testing means that a researcher forms a hypothesis and test the hypothesis using a series of quantitative or qualitative statistical methods to determine the statistical significance of the hypothesis being tested. The scientific method is based on experimentation to test a hypothesis .

Population geneticists usually do not test hypotheses. They make inferences about data based on Bayesian statistical inferences. They do not use statistical methods to determine the statistical significance of a hypothesis, they use statistics to describe data being reviewed by the researcher based on the beliefs the researcher already holds about the data being reviewed..

Population genetics is a type of Expost facto research. Expost facto research design is a quasi-experimental type of study examining how an independent variable, present prior to the research study, affects a dependent variable.

Whereas the subjects in experimental research are randomly selected, the participants in Expost facto research , are not randomly selected or assigned.The genome of the research subjects is examined to determine the haplotypes and haplogroups carried by the participants in the study.

In population genetics research the researcher uses the Bayesian inference method of statistical inference. The Bayesian statistical method, is a subjective research design/method that provides a rational method of updating the researcher's beliefs.

Since, the results of a Bayesian statistical analysis are a series of beliefs based on statistical inferences, the results can not stand alone. This is due to the reality, that any results, reported by a researcher are only a series of inferences based on the researcher's belief about a phenomena backed up by a series statistical results. If the results are published without corresponding evidence from archaeology, anthropology, linguistics and or craniometrics the inferences are pure conjecture, because they reflect the attitudes already held by the researcher, confirmed by data selected by the researcher to support his or her beliefs.

There is a sociological basis behind how a researcher interprets data. Sociological research indicates that there are unconscious cognitive structures within each individual. Cognitive structures that hold the idealistic view of members of the academe that determine how they perceive "reality". These structures are called doxa (Berlinerblau 1999).

Commenting on these schema Berlinerblau (1999) noted that "These types of theories share the assumption that human beings know things that they do not even know that they know; that they "possess" knowledge about the world which exists in some sort of cognitive substrate, beyond the realm of discourse" (p.106).Wacquant (1995) says that doxa is " a realm of implicit and unstated beliefs".

Given the research suggesting that doxa exist, support the view that some researchers allow their hatred of multiculturalism, ethnic prejudice and racism to define their discourse, teaching and writing about themes relating to groups " other" ,than their own cultural and ethnic group . Moreover, it suggest that when topics such as Eurasian and African haplogroups, Afrocentrism, African origins of the Dravidians and etc., is attacked by members of the academe, these academics are supported by the "establishment" without any reservation, or test of the validity of their claims. In fact, it appears that doxic assumptions relating to the validity of back migration of so-called Eurasian genes into Africa, recent African origin of Dravidians and Dravidian origin of the Indus Valley Civilization obviates critique of the academics that disparage these themes. Due to Doxa you can state a researcher's attitude toward a historical, genetic or anthropological concept and theorems without the statement being an ad hominem

Discussion

To evaluate research literature a student should know the varied research methods.A student evaluating a piece of population genetics' literature must understand that the researcher is conducting an expost facto method of research that does not involve hypotheses testing .Given the nature of Bayesian inferences, you cannot determine the validity and reliability of a piece of genetics research literature based on the statistical significance of the data. What you must do is look at the research article and ask yourself a series of questions regarding the article's validity and reliability.

To facilitate evaluation of genetics research literature I have created a check list: Checklist used to analyze a Population Genetics Papers, to evaluate research articles.

To use the Checklist you would perform the following task. The Evaluator should read the article twice. The first reading of the article is brief.

Next make a close reading of the article. The close read should involve the Evaluator in underlining key details in the article, while making annotations of important points in the text.

During the second reading of the text the Evaluator will assess the research article using **the Checklist used to analyze a Population Genetics Papers**.Since the Bayesian statistics used for the study will support the inferences of the Researcher the answers for the majority of the questions on the checklist will be yes.

The key question in determining the validity of the research will be question 17. If the researcher only has Bayesian statistical inferences supporting the research study , the inferences made in the research article , may not be representative of actual past population events.

I will use the Checklist to evaluate a recent Population genetics article. The paper is Chaubey and Endicott (2015). As mentioned earlier Bayesian statitistics, since they are based on the author's belief, will just about always support the author's inference. Below are my responses to thearticle placed on the Checklist. The evaluation of this article revealed the following responses:

1-3 is yes
4. No
5.yes
6. yes
7. no
8. no
9. yes
10. yes
11. yes
12. yes
13. no
14. no. No discussion of Southeast Asian and mainland Indian archaeology.
15. yes
16. no
17. No

Because the answer to Question 17, was no, demands that we check the archaeology literature to determine if the

Bayesian statistical inferences can find support from the craniometric, and archaeological record for SEA and India, or if the results and conclusion are based solely on the doxa of the researchers.

Claim that the Onge, a mainland Munda group only recently came to India circa 26kya. This would place them in India after the alledged settlement of India by the Aryans. They wrote: "of the Andaman-specific mtDNA lineage M31a1 around 26 ka, while the   ages of the diversification within M32 and M31a1 are estimated to fall within the Holocene, using whole-genome data in a Bayesian statistical setting (Barik et al.2008). Because mtDNA divergence is anticipated to predate population divergence, collectively these estimates suggest that the Andamans were settled less than ~26 ka and that differentiation between the ancestors of the Onge and Great Andamanese   commenced in the Terminal Pleistocene. Interestingly, this time frame is similar to the signal for population expansion found throughout ISEA (Guillot et al. this issue) and represents the time of topographic transition from the vast expanses of
Sundaland to the submerged Southeast Asian island chains of the Holocene.

In conclusion, we find no support for the settlement of the Andaman Islands by a population descending from the initial out-of-Africa migration of humans, or their immediate descendants in South Asia. It is clear that, overall, the Onge are more closely related to Southeast Asians than they are to present-day South Asians.

The similarity in proportions of the Onge genomes, attributed to the Melanesian, Malaysian (Jehai and Kensui), and South Asian ancestral components, combined with evidence for genetic drift, suggests that these constituent parts were present prior to their isolation from other parts of Southeast Asia".

Although this is the opinion of Chaubey and Endicott (2015), the Onge and other Munda populations were in India long before the Aryans. C Winters (2010) argues that Thangaraj et al using coalescence time and archaeological evidence illustrated that the TRMCA for mtDNA R8 which is found among Munda speakers have the following dates : R8 (41.7 kya), R8a (15.4 kya) and R8b (27.7 kya)13. The dating for mtDNA R8 indicates that this haplogroup and R7 are probably autochthonus to India.

The mtDNA of Munda speakers also includes deep rooted haplogroups from macrohaplogroup M. In addition to mtDNA haplogroup M2, we also find M58, M31, M6a2 and M42 among Munda speakers.

The Munda y-chromosome is O2a (M95). Kumar reports a coalescent rate of 65kya for Indian M953. There is a clear distinction of Indian Munda and Southeast Asian (SEA) Mon-Khmer speakers. The predominate SEA O clades are O3 and O1a. If SEA males had carried the y-chromosome O haplogroup to India there should be evidence of these clades among the Munda speakers—but they are nil8. On the otherhand, SEA males carry Indian y-chromosomes such as F,H, K2 (T) and etc8.

This indicates an early migration of Munda speakers to SEA. It suggest that Munda spread mtDNA R7 and y-chromosome haplogroup O to SEA.

Many Indians carry Munda haplogroups. The spread of Munda haplogroups are probably the result of conquest and intermarriage. The mythology of some Indian populations support this proposition. In other words, instead of the Munda originating in SEA, they probably migrated to the region from India.

Chaubey et al, based his conclusion on the research on Endicott et al (2006).Endicott et al (2006) argue that without comprehensive data from Myanmar it is not possible to identify whether the Andaman M31a1 arrived from India or if the Indian M31a2 came from South-East Asia. But either scenario casts serious doubts on the concept that the Andaman Islands were settled at the time of the migrations out of Africa carrying the current Eurasian mtDNA diversity".

Endicott et al (2006), admit that their conclusions should be preliminary because: "Without comprehensive data from Myanmar it is not possible to identify whether the Andaman M31a1 arrived from India or if the Indian M31a2 came from South-East Asia. But either scenario casts serious doubts on the concept that the Andaman Islands were settled at the time of the migrations out of Africa carrying the current Eurasian mtDNA diversity".

It is obvious that Endicott et al (2006) , could not answer this question because they did not know much about Southeast Asian history.

If they knew the archaeology of Southeast Asia they would have been able to answer this question. They would have known that the Dravidians who carry M31a2 probably carry the haplogroup as a result of migration of Dravidian back to South India from Myanmar. Winters (2010), I explain0 that, many Dravidian speakers in India formerly lived in Southeast Asia.Formerly intimate relations existed between South Indians and Southeast Asian people (Kanakasahai, 1966). The Tamilian form of Saivism is known as Agamas, the esoteric and ritualistic parts of Agama is non-Vedic (not of Indo-European origin). Agama was also the Southeast Asian form of Hinduism (Winters,1985).

The Proto-Tamil speakers in Central Asia and China were called the Yakshas in Indian literature (Yuehchih by the Chinese) and Kosars (Kushana in Chinese literature). They were forced from China due to first the classical Mongoloids who founded Shang-Yin , then the Zhou and succeeding mongoloid Chinese and Thai populations that invaded Indo-China . This forced the Proto-Tamil speaking Kosars and Yakshas to later invade southern India in search of a new homeland in addition to Southeast Asia (Winters, 2011). In Southeast Asia Dravidian speakers probably encountered Proto-Andamanese carrying M31 and M32 who may have been the original settlers of the area.

The archaeological, and genetic evidence indicate that Dravidian speakers lived in Southeast Asia (Kanakasabhai, 1966; Winters, 1985) . It indicates that the first civilizations in Southeast Asia were founded by Dravidian speakers (Kanakasabhai,1966)..

The Khmer introduced various aspects of civilization in this region which precede the advent of the Thai speakers into this region. Upon their arrival in Indo-China ,the Thai-

Vietnamese people conquered the blacks learned their culture and continued to perpetuate the same cultural traits (Winters,1985).Thusly, we see that both the Vietnamese and Thai peoples learned their culture, architecture, religion and writing from the Khmers and other Indo-African people.

While the Dravidians lived in Southeast they probably mated with the inhabitants related to the Andamanese (Winters,2011). This mating pattern probably led to M31a2 entering the Dravidian gene pool when the Kamboja settled in Sengal and South India (Kanakasabhai, 1966).

Conclusion

In summary we can reject the research of Gyaneshwer Chaubey and Phillip Endicott, based on question 17 of the Checklist used to analyze a Population Genetics Papers, because it is unreliable and lacks validity because the researchers failed to study the archaeology and history of SEA. If they had, they would have known that Dravidian speakers formerly lived in SEA, until the advance of the Classical mongoloid people 2.5kya.

In summary, the validity and reliability of a piece of genetics research literature does not demand the Evaluator of a piece of literature to provide counter evidence all they need to do is evaluate the research using the checklist (see Appendix). If the answer to most of these questions is no, the research is unreliable and lacks any validity.

The key question on the checklist is question 17. To confirm the validity of the archaeological, craniometric and etc., data , the Evaluator should be knowledgeable about the archaeology of the area where the population movement has been inferred to have taken place. In this way you can determine if the Bayesian inferences correspond to the archaeological, craniometric, and linguistic data associated with the geographical area where the population movement is alleged to have occured .

The major problem with most genetics literature which invalidates the research dealing with ancient population movements is that it is not supported by the ancient DNA, archaeological and/ or craniometric data. This is why many of theories about the ancient populations of Europe and alledged back migrations are usually over turned once researchers examine the ancient DNA.

Reference:
Ackland G J , Markus Signitzer, Kevin Stratford,and Morrel H. Cohen.(2007).Cultural hitchhiking on the wave of advance of beneficial technologies PNAS 104 (21) 8714-8719; published ahead of print May 16, 2007, doi:10.1073/pnas.0702469104. Retrieved 2/6/2015 at : http://www.pnas.org/content/104/21/8714.full
Berlinerblau, J. (1999). Heresy in the University: The Black Athena Controversy and the Responsibilities of American Intellectuals .Rutgers University Press.
Chaubey, G and Phillip Endicott. (2013)The Andaman Islanders in a Regional Genetic Context: Reexamining the Evidence for an Early Peopling of the Archipelago from South Asia. Retrieved 3/6/2015 at: http://digitalcommons.wayne.edu/cgi/viewcontent.cgi?article=2055&context=humbiol
Endicott P, Metspalu M, Stringer C, Macaulay V, Cooper

A, et al. (2006) Multiplexed SNP Typing of Ancient DNA Clarifies the Origin of Andaman mtDNA Haplogroups amongst South Asian Tribal Populations. PLoS ONE 1(1): e81.
http://journals.plos.org/plosone/article?id=10.1371/journ al.pone.0000081

Kanakasabhai,V.(1966).   The Tamil Eighteen Hundred Years ago .

Renfrew,   C.   (2001).From   molecular   genetics   to archaeogenetics   PNAS   98   (9)   4830-4832; doi:10.1073/pnas.091084198.
http://www.pnas.org/content/98/9/4830.full

Winters, C. (1985).   "The Far Eastern Origin of the Dravidians", Journal of Tamil Studies, pp.66-92.

Winters.C. (2010). Munda Speakers are the Oldest Population in India. The Internet Journal of Biological Anthropology.   2010   Volume   4   Number   2,
https://ispub.com/IJBA/4/2/5591

Winters, C. 2011. Comment : A back migration from Southeast Asia accounts for M31a2 in South India,
http://www.plosone.org/annotation/listThread.action?roo t=609

Appendix
Checklist used to analyze a Population Genetics Papers

Answer the following questions relating to this research article below, or on a separate sheet of paper.

1. What was the rationale for the study, that is, what led up to it? Yes on page___, paragraph____, lines_____ No_____

2. Why do the authors believe that this problem is significant? Yes on page___, paragraph____, lines_____ No_____

3. What was the purpose of the study, that is, what did it intend to accomplish? Yes on page___, paragraph____, lines_____ No_____

4. What was the hypothesis of the study? Yes on page___, paragraph____, lines_____ No_____

5. What were the participant's major characteristics? Yes on page___, paragraph____, lines_____ No_____

6. Does the review of literature indicate previous research in the area associated with the article? Yes on page___, paragraph____, lines_____ No_____

7. What type of study is reported in this article? Yes on page___, paragraph____, lines_____ No_____

8. Was the sample randomly selected? Yes on page___, paragraph____, lines_____ No_____

9. What was the instrument? Yes on page___, paragraph____, lines_____ No_____

10. What were the major steps involved in the treatment? Yes on page___, paragraph____, lines_____ No_____

11. How were the variables tested? Yes on page___, paragraph____, lines_____ No_____

12. According to the author(s) how successful was the treatment? Yes on page___, paragraph____, lines_____ No_____

13. What factors could equally account for the student tests results? Yes on page___, paragraph____, lines_____ No_____

14. What problems, if any, do you detect in the study? Yes on page___, paragraph____, lines_____ No_____

15. Do the results of analysis agree with the author's objectives and expectations? Yes on page___, paragraph____, lines_____ No_____

16. What other interpretations could be made from the data? Yes on page___, paragraph____, lines_____ No_____

17. Is there archaeological, craniometric and or linguistic evidence that supports the research findings yes on page___, paragraph____, lines_____ No_____

ABOUT THE AUTHOR

Dr. Clyde Winters is an Educator and Anthropologist. He is the Director of the Uthman dan Fodio Institute; an Educational and Anthropological Research Institute in Chicago, Illinois founded by Dr. Winters.

Dr. Winters taught Education and Linguistics at Saint Xavier University-Chicago. He taught Social Studies and Special Education in the Chicago Public Schools for 43 years. At CPS Dr. Winters helped write the Social Science and Common Core State Standards for the CPS.

In addition Dr. Winters has deciphered the Olmec, Meroitic and Indus Valley writing. He has published over 100 articles in anthropology, linguistics, population genetics and education.

Printed in Great Britain
by Amazon